MANUFACTURING COST ESTIMATING

Phillip F. Ostwald
Editor

Published by:
Society of Manufacturing Engineers
Marketing Services Dept.
One SME Drive
P.O. Box 930
Dearborn, Michigan 48128

Manufacturing Cost Estimating

Copyright, 1980 by the
Society of Manufacturing Engineers
Dearborn, Michigan 48128

First Edition

All rights reserved, including those of translation. This book, or parts thereof, may not be reproduced in any form without permission of the copyright owners. The Society does not, by publication of data in this book, ensure to anyone the use of such data against liability of any kind, including infringement of any patent. Publication of any data in this book does not constitute a recommendation of any patent or proprietary right that may be involved.

Library of Congress Catalog Card Number: 79-67648

International Standard Book Number: 0-87263-053-6

Manufactured in the United States of America

SME wishes to express its acknowledgement and appreciation to the following publications for supplying the various articles reprinted within contents of this book

AACE Bulletin
American Association of Cost Engineers
308 Monongahela Building
Morgantown, WV 26505

American Machinist
McGraw-Hill, Inc.
1221 Avenue of the Americas
New York, NY 10020

Appliance Engineer
Dana Chase Publications, Inc.
York Street at Park Avenue
Elmhurst, Ill. 60126

Industrial Engineering
American Institute of Industrial Engineers, Inc.
25 Technology Park/Atlanta
Norcross, GA 30092

Machine and Tool Blue Book
Hitchcock Publishing Company
Hitchcock Building
Wheaton, Illinois 60187

Management Accounting
National Association of Accountants
919 Third Avenue
New York, NY 10022

Manufacturing Engineering
Society of Manufacturing Engineers
One SME Drive
P.O. Box 930
Dearborn, MI 48128

Grateful acknowledgement is also expressed to:

American Association of Cost Engineers
308 Monongahela Building
Morgantown, WV 26505

American Institute of Industrial Engineers, Inc.
25 Technology Park/Atlanta
Norcross, GA 30092

Prentice Hall, Inc.
Englewood Cliffs, NJ 07632

Society of Plastic Engineers
656 West Putnam Avenue
Greenwich, CT 06830

PREFACE

Because of specialization, many engineers, technologists and business managers are, collectively, practicing the art of cost estimating. Professionals are engaged in this work even though their job titles are not "Manufacturing Cost Estimators." The prominence of the cost estimating specialty is now an accepted fact in manufacturing. This books should stimulate the understanding and development of those engineers, technologists and business managers who deal with cost estimating problems.

Economic prosperity and survival have always been the concern of the American manufacturing scene from its early origins of the cottage trade to the mid-1800's Industrial Revolution and to our current level of technology. In manufacturing, the estimate and measurement of cost for operations and products is closely connected to survival and long-term prosperity.

Most organizations treat the estimate value as the true value. Avoiding blunders and errors of policy and information are crucial to an acceptable quality of estimating, particularly today.

The first chapter of this book presents fundamentals of manufacturing cost estimating. Direct labor and direct material costs are the gist for the estimate and it is clear that perception of these topics is necessary for advanced understanding of cost estimating.

Computer estimating is vital for larger and more complex estimating. It should be understood that the computer retrieves, calculates, recombines and stores information as a data processing tool. As estimating depends upon experience and judgment of a person, the computer does little estimating. Estimating cannot be a mechanical or an electronic exercise alone. However, the organization of business and estimating data has progressed far beyond the estimator's little black book. Chapter Two describes recent innovations.

There are over 100,000 firms in the United States that manufacture a product. Each of these firms has a method for the collection, arrangement and display of cost data. "Proforma," a Latin term meaning "for the sake of form," deals with the methods of making a comprehensive estimate. Management is, of course, interested in knowing the steps in estimating, from where the procedure starts to the important bottom line. Procedures, recap sheets, and summary are other terms that each company uses to title its proforma.

For some industries, especially those that produce industrial capital goods (airplanes, machine tools, etc.), the learning curve is a codified tool that describes the decrease in unit cost or time. The concept is increasingly being adopted by commercial firms and Chapter Four provides theory and practice.

The issue of tool estimating remains as always: (1) how to find the cost or time, and (2) how to amortize the cost over the lot, yearly or model quantity. These points are discussed by Chapter Five.

What are the new ideas that are useful to make better estimates? Cost control is not an isolated function; it sends performance signals to cost estimating as measured against time and cost. These and other thoughts emerge from Chapter Six. Estimating has become a vital part of modern American industrial management. The care and feeding of this specialization for its professional improvement is described in Chapter Seven.

I wish to express my gratitude to the authors whose works appear in this volume. I also express my thanks to the publications who generously allowed us to use their material. They are: *AACE Bulletin, American Machinist, Appliance Engineer, Industrial Engineering, Machine and Tool Blue Book, Management Accounting,* and *Manufacturing Engineering.* Special acknowledgement should go to the American Association of Cost Engineers, the American Institute of Industrial Engineers Inc., Prentice-Hall Inc., and the Society of Plastic Engineers for their contributions to the development of this book. My thanks is also extended to Bob King of the SME Marketing Services Department for his efforts in producing this volume.

Phillip F. Ostwald
Professor of Mechanical and Industrial Engineering
Department of Mechanical Engineering
University of Colorado
Boulder, Colorado
Editor

SME

The informative volumes of the Manufacturing Update Series are part of the Society of Manufacturing Engineers' effort to keep its members better informed on the latest trends and developments in engineering.

With 50,000 members, SME provides a common ground for engineers and managers to share ideas, information and accomplishments.

An overwhelming mass of available information requires engineers to be concerned about keeping up-to-date, in other words, continuing education. An SME Member can take advantage of numerous opportunities, in addition to the books of the Manufacturing Update Series, to fulfill his continuing educational goals. These opportunities include:

- Chapter programs through the over 200 chapters which provide SME members with a foundation for involvement and participation.
- Educational programs including seminars, clinics, programmed learning courses and videotapes.
- Conferences and expositions which enable engineers to see, compare, and consider the newest manufacturing equipment and technology.
- Publications including Manufacturing Engineering, the SME Newsletter, Technical Digest and a wide variety of books including the Tool and Manufacturing Engineers Handbook.
- SME's Manufacturing Engineering Certification Institute formally recognizes manufacturing engineers and technologists for their technical expertise and knowledge acquired through years of experience.

In addition, the society works continuously with the American National Standards Institute, the International Standards Organization and other organizations to establish the highest possible standards in the field.

SME members have discovered that their membership broadens their knowledge throughout their career.

In a very real sense, it makes SME the leader in disseminating and publishing technical information for the manufacturing engineer.

TABLE OF CONTENTS

CHAPTERS

1 FUNDAMENTALS FOR MANUFACTURING COST ESTIMATING

AN INTRODUCTION TO COST ESTIMATING
By *A. N. Paul*
From "Preparing And Controlling the Cost Estimate" 3

JOB ORDER COSTING AND DEFINITIONS
By *Gordon F. Gantz*
Reprinted by special permission from the Cost Engineers' Notebook. Copyright © 1976 by the American Association of Cost Engineers, 308 Monongahela Building, Morgantown, WV 26505. All rights reserved 9

DEVELOPMENT AND USE OF MACHINABILITY DATA FOR PROCESS PLANNING OPTIMIZATION
By *Vijay A. Tipnis,, Michael Field* and *Moshe Y. Friedman*
Presented at the CAD/CAM III Conference, February, 1975 15

A PRIMER ON MATERIAL ESTIMATING
By *Phillip F. Ostwald* and *Patrick J. Toole*
Reprinted by special permission from (1977) Transactions of the American Association of Cost Engineers. Copyright © (1977) by American Association of Cost Engineers, 308 Monongahela Building, Morgantown, WV 26505 29

FOUR SLIDES, PART 6—ESTIMATING YOUR COSTS
Reprinted from Manufacturing Engineer, September, 1978 33

PSQL AN ECONOMIC CRITERION FOR MINIMIZING OVERALL INSPECTION AND REPAIR COST
By *Michael Z. Freuhwirth*
Reprinted by special permission from the (August, 1975) AACE Bulletin. Copyright © (1975) by American Association of Cost Engineers, 308 Monongahela Building, Morgantown, WV 26505 35

ESTIMATING OF MANUFACTURING JOINT COSTS
By *Adam Malolepszy* and *Phillip F. Ostwald*
Prepared for MANUFACTURING COST ESTIMATING 38

COMPUTERIZED STANDARD DATA A POWERFUL NEW TOOL FOR MANUFACTURING ENGINEERING
By *Leroy H. Lindgren* and *Romeyn D. Murphy*
Presented at the CAD/CAM Conference, February, 1972 45

STANDARD TIME DATA
By *E. A. Cyrol*
Reprinted from Machine and Tool Blue Book, July, 1961 60

2 COMPUTER AIDED ESTIMATING

MACHINE JOB SHOP PREPARES QUOTATIONS BY COMPUTER
By *Lloyd D. Doney*
Reprinted with permission from Industrial Engineering Magazine (August, 1971). Copyright American Institute of Industrial Engineers, Inc., 25 Technology Park/Atlanta, Norcross, GA 30092 71

JOB SHOP COST ESTIMATING USING COMPUTER TECHNIQUES
By *Donald G. Radke*
Reprinted by special permission from (1977) Transactions of the American Association of Cost Engineers. Copyright © (1977) by American Association of Cost Engineers, 308 Monongahela Building, Morgantown, WV 26505 76

THE USE OF MINICOMPUTERS IN COST ENGINEERING
By *Raymond P. Wenig*
Reprinted by special permission from (1976) Transactions of the American Association of Cost Engineers. Copyright © (1976) by the American Association of Cost Engineers, 308 Monongahela Building, Morgantown, WV 26505 .... **80**

COST ESTIMATING AND PRODUCTION PLANNING AT THE PRELIMINARY DESIGN STAGE
By *Daniel E. Strohecker*
Presented at the WESTEC Engineering Conference and Tool Exposition, March, 1970 .... **90**

COMPUTER ESTIMATION OF MOLDING AND TOOLING COSTS FOR WOODGRAIN PLASTIC PARTS
By *John E. Johnson*
Printed with permission of the Society of Plastics Engineering **104**

3 PRO FORMA

PRO FORMA IN MANUFACTURING COST ESTIMATING
By *Phillip F. Ostwald*
Prepared for MANUFACTURING COST ESTIMATING **111**

ESTIMATING LABOR AND BURDEN RATES
By *J. G. Margets*
Reprinted from Manufacturing Engineering, January, 1972 .... **123**

HOW TO SIMPLIFY COST ESTIMATES
By *Samuel L. Young*
Reprinted from Manufacturing Engineering, February, 1972 .... **126**

VARIABLE COST ESTIMATING
By *Phillip F. Ostwald* and *Gary E. White*
Reprinted with permission. Copyright American Institute of Industrial Engineers, Inc., 25 Technology Park/Atlanta, Norcross, GA 30092 .... **129**

PREPARING ESTIMATES FOR DEVELOPMENTAL WORK
By *Paul D. Fowler*
Presented at the WESTEC Engineering Conference, March, 1974 .... **142**

ESTIMATING PRODUCT ENGINEERING COSTS
By *Phillip F. Ostwald*
Reprinted from Appliance Engineer magazine, copyright Dana Chase Publications, Inc., 1972 .... **150**

4 LEARNING CURVE

LEARNING CURVE
From the book Cost Estimating for Engineering and Management
By *Phillip F. Ostwald*
Copyright 1974, Prentice Hall, Inc. **157**

LEARNING CURVES IN MANUFACTURING COST ESTIMATES
By *Raymond B. Jordan*
Presented at the Economics of Material Removal Seminar, February, 1970 .... **167**

5 TOOL ESTIMATING

HOW TO ESTIMATE DIES, JIGS, AND FIXTURES FROM A PART PRINT
By *Leonard Nelson*
Reprinted from American Machinist, September 4, 1961 © McGraw-Hill, Inc. **185**

HOW DO YOU HANDLE TOOLING COSTS?
By *Phillip F. Ostwald* and *Patrick J. Toole*
Reprinted with permission from Industrial Engineering Magazine, (November, 1975). Copyright American Institute of Industrial Engineers, Inc., 25 Technology Park/Atlanta, Norcross, GA 30092 .... **197**

6 COST CONTROL USING ESTIMATES

WHAT WILL YOU PAY FOR ENERGY NEXT YEAR?
By *Phillip F. Ostwald*
Reprinted with permission from Industrial Engineering Magazine, (January, 1979). Copyright American Institute of Industrial Engineers, Inc., 25 Technology Park/Atlanta, Norcross, GA 30092 .... **205**

PROCEDURES FOR CONTROLLING MANUFACTURING COSTS IN THE DURABLE GOODS INDUSTRIES
By *H. K. Von Kaas*
Reprinted by special permission from (1977) Transactions of the American Association of Cost Engineers. Copyright © (1977) by American Association of Cost Engineers, 308 Monongahela Building, Morgantown, WV 26505 .... **209**

LIFE CYCLE COSTING
By *Gary E. White* and *Phillip F. Ostwald*
Reprinted from Management Accounting, January, 1976 .... **214**

YOUR COST SYSTEM: A MANAGEMENT OR ACCOUNTING TOOL?
By *Robert L. Dewelt*
Reprinted from Manufacturing Engineering, May, 1977 .... **217**

COST ESTIMATING WITH UNCERTAINTY
By *Jim D. Burch*
Reprinted with permission from Industrial Engineering Magazine, (March, 1975). Copyright American Institute of Industrial Engineers, Inc., 25 Technology Park/Atlanta, Norcross, GA 30092 .... **223**

7 MANAGING A COST ESTIMATING TEAM

TOWARDS THE COST ESTIMATING SPECIALTY
By *Phillip F. Ostwald* **229**

TOP MANAGEMENT AND THE ESTIMATING FUNCTION
By *Robert C. Berman*
Presented at the Manufacturing Management Conference, May, 1976 .... **237**

FABRICATION AND ASSEMBLY COST ESTIMATING
By *Arnold M. Kriegler*
Presented at the WESTEC Engineering Conference and Tool Exposition, March, 1970 .... **245**

INDEX **253**

CHAPTER 1
FUNDAMENTALS FOR MANUFACTURING COST ESTIMATING

An Introduction to Cost Estimating

By A. N. Paul

A. What is a cost estimate?

   A cost estimate is a <u>forecast</u> of expenses <u>that may</u> be incurred in the manufacturing of a product. The estimate may only include the factory costs but usually not the selling price which must include the burden, secondaries, handling, warehouse or storage, profits and other elements that are added to the factory cost to obtain the selling price.

   Cost estimating is different from cost accounting in that it is before the fact and thus a forecast, opinion, appraisal, educated guess, or a scientific analysis of all the factors entering into the production of a proposed product. Cost accounting is after the fact recording, analyzing and interpreting the historical data of dollars and hours that were used to produce a particular product at a given time under conditions existing at the time. Cost accounting tells management the effectiveness of production while cost estimating gives information as to probable costs in dollars or hours for making parts, products, or using certain processes or particular operations.

B. The purpose and uses of cost estimates.

   1. To provide executive information to be used in establishing the selling price for quotation, bidding, or evaluating contracts. The purpose is to insure that bids if accepted will provide a reasonable profit for the company and to price quotations low enough to enhance the chances of the company to be awarded the contract.

   2. Ascertain whether a proposed product can be made and marketed at a profit considering existing prices and future competition.

   3. To be used as a make or buy decision thus determining whether parts and assemblies can be made or purchased from a vendor more cheaply.

   4. To determine how much may be invested in tools and equipment to produce a product or component by one process as compared to another method.

   5. To determine the best and most economical method, process, or material for the manufacturing of a product.

   6. To be used as a basis for a cost reduction program thus showing savings that are or may be made by the changing of methods, and processes; or the application of value analysis techniques.

7. To predetermine standards of production performance which may be used in the control of costs of operations for a product at the start of production.

8. To predict the effect of volume changes on future profits by the introduction of automation, mechanization, or other improvements suitable to mass production.

9. In the case of new products the estimate details will include the first formal process planning which may later become the basis for the following:

   a. Establishing personnel requirements to meet future work plans.
   b. Predict material needs over the length of the contract.
   c. Set the overall schedule or time table for meeting company goals.
   d. Specifying the equipment, machines, and facilities for production of the proposed product in the time and quantities required.

10. An estimate is the dollar answer to a bid request to supply an item. The specified item may be anything from raw material to an industrial complex, hardware or paperwork, services or research and development, construction, etc.

C. The types of cost estimates are as follows:

1. The quick response estimate: This is usually a one or two man effort for an hour or two but seldom more than a day in which the estimator gives his top management a figure of first order of magnitude, (at the best not expected to be accurate to more than + or - 30%) estimate of an overall cost for a product. On the basis of this report a decision may be made to either:

   a. Forget the project as it has little hope of profits.
   b. Go ahead with detail planning at full speed with an all out effort.
   c. Look into the matter further with a more detailed but preliminary effort because management is interested but needs more facts before commitment.

2. Preliminary estimate: This is based on assumptions, areas of incomplete data, designs and specifications which are not yet complete or frozen, and using judgment of experienced operators or specialists who have had long and wide industrial backgrounds of similar jobs to supply values for missing data. The accuracy of either of these first two kinds of estimates are only as good as the experience of the estimators but they are fast and

practically no cost is involved in producing the estimate. Oftentimes one of the estimators or his immediate superior has enough of his own dollars in the company so that he in a sense is answerable to himself or partner for his decision. In a very small business and for simple products, these quick off the cuff estimates have served their purpose. However, in large companies dealing with sophisticated systems such estimates are dangerous and should be avoided.

3. Past average cost and use of previous final cost estimates: This system uses past actual costs of somewhat similar products or previous estimates of comparable parts. While this type is in more detail it is still a short cut in the preparation of estimates and usually is done by clerical procedures referring to file records which may or may not be applicable to the problem. It is usually necessary to make substitutions, or revisions in the past data and former estimates to account for new materials or the use of new or improved methods. If the product is one which has been made in the past the learning curves should reduce the new cost to approximately 80% each time the number of parts made are doubled provided the same people and machines are used. Historical data may contain many irregularities as to time, cost, standards, bad luck, bad weather or company catastrophe which would make past data invalid.

4. Composite rates or overall ratios: This type estimate combines some or all of the cost elements into a single rate or ratio such as so many $/lb.; $/sq. ft./ $/cu. ft. or so many $ per any unit of a particular product. These ratios are sometimes called rule-of-thumb formulas for rough estimating. All of these must be used with considerable judgment and considerations as to how the job at hand differs from the normal on which the ratio was established. Some of these rule-of-thumb formulas follow:

   a. DL = 50% of direct cost = 25% of manufacturing cost
      (DL = direct labor)
   b. OH = 2 x DL = DL + (1.5 x direct material) = 3 x direct material
   c. Building costs - 20 $/sq. ft.
   d. 1 1/2 x direct material = OH = DL = 50% direct costs = 25% of manufacturing cost
   e. Sales and distribution costs and administration = 1/2 the retail price
   f. Retail price = 8 x raw material cost x a factor for complications of part or the labor content

   The uses and limitations of these formula will be discussed in a later lesson.

5. Predetermined cost for final cost estimate: This detailed estimate is most accurate. This estimate is based on the complete data about the product, the use of proven standards over a long period, the coordination of information from all departments involved in making the product, complete processing plans for each detail part of the product, tooling designs and their cost estimates, and the use of predetermined time standards and synthetic times for each operation in all processes.

D. The elements of a cost estimate:

All expense items to a company must be considered in the established selling price. The general and administrative (G & A) costs as well as the profit must be added to the manufacturing costs. Since the accounting department usually establishes G & A and management may establish the profit only the manufacturing cost is of prime interest to the cost estimator. If the formula given under the composite rate estimating were completely valid the following diagram would show the proportionate weight in each section and its share of the list price.

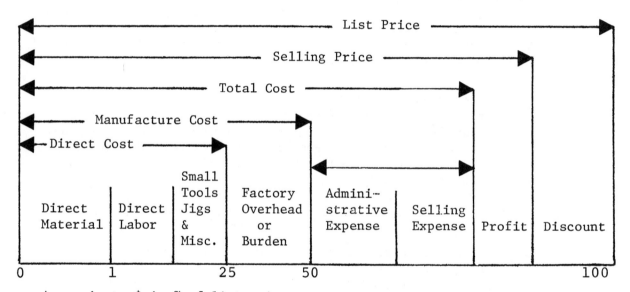

Approximate $ in % of list price

Note that the list price or retail price is equal to many increments; only 1/2 of which is the manufacturing or product cost that is the subject of this course and which is of utmost importance to the cost estimator. It may also be noted that since all increments in the list price are obtained in accounting by applying ratios to the material and or labor, any reduction in labor or material that the estimator can obtain via such techniques as value analysis reduces the selling price or increase profits by from 4 to 8 times the saving.

To increase the profit the selling price must be increased or the various cost increments must be reduced. The process planner, product designer, tool designer, industrial engineer, and or labor can materially increase or decrease the profits of a company, by the way and efficiency in which they do their job.

E. Manufacturing cost estimating:

As shown in the diagram on page 6, the manufacturing cost is composed of the direct material; small tools, jigs and fixtures, direct labor, and the factory (OH) overhead sometimes called burden or manufacturing expense.

Direct material is a product ingredient which becomes a part of the finished product and can readily and inexpensively be traced to the finished product as so many units per product made. Thus the number of parts made is directly related to the amount of material used.

Direct labor is that work which is actually used in producing a finished article and is readily charged to it. DL changes the material form or part as it goes through his station. A person doing direct labor usually touches the part or material with his hands, a tool, or some other material such as paint.

Small tools, jigs and fixtures are the accessories used in production processes which are expendable and charged to the product. Cutting and cleaning fluids or other process materials would be included in this element.

Burden, factory overhead, or manufacturing expense consists of all indirect labor, materials and other indirect manufacturing expenses that cannot be readily allocated to a product as so much used per part made. Facilities, equipment, power, air and utilities as well as manufacturing services such as maintenance are a part of the cost of this element. Much of the burden is fixed and thus does not vary with the number of parts made per day.

F. Factors in cost estimating:

There are numerous factors that may enter into the estimating procedures and basic data which will cause deviations, variations, and errors in the final answer. Pitfalls to be avoided will be covered in Lessons 3 and 4. Sources of accurate basic data will be covered in Lesson 3. The following list some factors that may invalidate the estimate.

1. Under estimates may result in company losses and they may be due in part to:

   a. Higher than anticipated labor or materials cost (labor troubles, strikes, slow down, sickness of many key operators, and/or bad materials).

7

b. Incomplete design information at the time of the estimate or design changes introduced after the estimate is made. Changes requested by the customer should be payed for at a reasonable cost fee basis.
c. Unexpected delays in receiving the order, instructions, raw materials, etc.
d. Unexpected machine, processing, or assembly problems.
e. Errors in estimating such as computations, omissions of cost items, etc.

2. Higher estimates may be due to the following:

   a. Poor standard data or higher than actual basic data prices of material.
   b. Cost reduction practices introduced after estimate (value analysis, etc.).
   c. Improved processing methods (revolutionary new ways of making parts).
   d. Errors by the estimators and clerks preparing the data (no learning curve used).

3. High estimates may be as serious as the under estimate as they may lead to:

   a. Overbidding and rejection of proposals, in either case a loss of work.
   b. Loss of customer good will, possible loss of future business.
   c. Waste in design, fabrication, and the use of materials.
   d. Managements loss of faith in their estimators (perhaps loss of your job).

4. Estimating errors result from any of the following common variables:

   a. Fluctuations in the basic costs of materials or labor which are related to inflation and deflation of the complete economy.
   b. Equipment, tools, jigs, fixtures, or other processing costs may vary.
   c. Price breaks due to volume purchases of material or purchased parts.
   d. Revision of organization changing the burden rate either up or down.
   e. Major changes in the allocation of other manufacturing expense such as rents on buildings or facilities, taxes, and cost of facilities and utilities.
   f. The errors due to the estimator may be pure human mistakes but are more apt to be his lack of knowledge of materials, processes, labor, and/or his omission of some cost items from the estimate.

AMERICAN ASSOCIATION OF COST ENGINEERS

COST ENGINEERS' NOTEBOOK

Reprinted by special permission from the Cost Engineers' Notebook. Copyright © (1976) by the American Association of Cost Engineers, 308 Monongahela Building, Morgantown, WV 26505. All rights reserved.

JOB ORDER COSTING AND DEFINITIONS*
by Gordon F. Gantz, Wisconsin Section

Manufacturing costs include direct material, direct labor and factory overhead to arrive at <u>total factory cost</u>.

<u>Direct Material</u> is all material which is an integral part of and identifiable with the finished product, e.g., steel plate in trolley frames; insulated copper wire in wiring. Weld rod, for example, is considered indirect material because of the impracticality of tracing it to specific physical units of product.

<u>Direct Labor</u> is all labor which is easily related to specific products; for example, labor of machine operators and assemblers. Labor of material handlers and plant guards is considered indirect because of the difficulty of tracing such to specific physical units.

<u>Factory Overhead (or Burden)</u> includes all manufacturing costs except direct material and direct labor charged directly to individual job orders. Examples of overhead expenses would be: supplies, indirect labor (supervision, materials handlers), depreciation, electricity, fuel, insurance, property taxes.

1) Direct labor is charged to the job by knowing the total time (hours) worked ("Time On", "Time Off") on each job. Generally, factory overhead is charged to the job in proportion to the direct labor on the job and set up on the basis of departmental cost centers.

2) Since

   a) the relationship between total direct labor and total overhead cannot be determined precisely until the end of an accounting period;
   b) the relationship varies from period to period;
   c) because of a need for timeliness, overhead is applied by a predetermined rate per hour and is likely to be under or over-absorbed at the end of the accounting period.

The concept of <u>contributed value</u> is one to be mentioned at this point. Contributed value is factory cost minus direct material cost and is commonly used as a measure of the effectiveness of a manufacturing operation. In a sense a company has no control over the cost of direct materials it purchases, but does have control of all of the costs after the purchase of that direct material, and so may find it useful to know what these contributed costs are and how much they vary from time to time.

*Definitions are an expansion of ANSI Z94 as regards the manufacturing cost components (ANSI Z94—Cost Engineering Glossary of Technical Terms at present under review. These terms are included in this Notebook under Index No. AA-4,000).

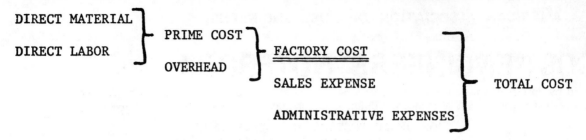

Fig. 1. Development of the Total Cost of a Product.

The cost terms here defined are the elements within a Cost Engineering Project, the purpose of which was:

to establish an accurate pricing base for new product sales and

for controlling current operation costs.

It should be further noted that on this project the measure of direct labor activity or base used for the application of overhead expenses were reported "earned-productive-hours" and not the usually reported "clocked" labor hours.

Earned productive hours are recorded for pay purposes where an incentive system exists. Thus, a job is rated in standard hours and the man may earn "more" dollars in a day by producing the job in shorter actual (clocked) hours. The advantages of the "earned" hours base is that it is the same for each product unit whereas the actual hours for a product unit varies with the man's performance (efficiency) on that particular job. Having a stable hour base the "earned" hour system simplifies variance determination.

DEFINITIONS OF JOB SHOP COSTS

1. Job Costs - costs for parts produced or operations performed on a job following the sales order.

   Examples of Job Parts - are those made to purchaser's requirements and cannot be produced before the sales order; e.g., overhead crane girders because crane span is dependent on purchaser's building width and needs cannot be known before the order.

   Job Operations are typically - assembly, wiring, painting, loading, fabrication, and welding.

2. Stock Ordered Costs - are costs for products produced on a job in advance of a sales order in anticipation of specific customer requirements.

   Examples of Stock Parts Costs - are structural weldments, machinery, most electrical parts and assemblies (motors, brakes, control, etc.).

3. Operating Cost Control - operating costs may be controlled by means of a cost standard which identifies what a part or operation should cost; by comparing actual (reported) costs with these cost standards to establish variances and finally from an analysis of these variances take corrective action to achieve the desired goal. (Corrective action, though usually taken to reduce the variance, may include modification of the standard.)

4. <u>Elements of Reported Costs</u> - reported costs are the summation of Day Work Hours and Earned Standard Hours (where an incentive pay system exists) reported against an order. The total earned standard hours and day work hours reported against a given production order are called the <u>Earned Productive Hours</u> (EPH). (See Paragraphs 5 and 6 below for definitions of Day Work Hours (DWH) and Earned Standard Hours (E)).

   When the job order includes the use of stock parts (made in advance of the job order), ordinary accounting convention for reporting the cost of "this" job order would be the summation of stock costs and job costs; e.g.,

Stock Costs	(Ss)	=	Stk. Hrs.	$Material	+	$Labor	+	$Burd.	= $FC
Job Costs	(E)	=	Job Hrs.	$Material	+	$Labor	+	$Burd.	= $FC
REPORTED COSTS	(R)		Hrs.	$M	+	$L	+	$B	= $FC*

   NOTE: The stock part standard cost is used and, thus, variances on stock parts are controlled in a performance - variance analysis system separately from job costs. Thus, in general, there are three situations for Reported Costs.

   a) All work produced to Stock -
   b) All work produced on the job (MOJ)
   c) The job work added to parts made for stock.

   *$FC means Factory Cost Dollars and is the summation of direct material dollars ($M), direct labor dollars ($L), and all indirect overhead or burden ($B) (material, labor and expenses) attributable to the product.

5. <u>Day Work Hours</u> (DWH) (Direct Labor) are the actual job hours reported on production order numbers for the pay period for which no incentive rates are provided. Also, for non-incentive systems, daywork is the time worked by direct labor.

6. <u>Earned Standard Hours</u> - are the hours reported by multiplying the standard hour incentive <u>rate</u> per unit by the total units produced.

7. <u>Attendance Hours (C)</u> - are the regular straight time (clocked) hours a man is at work as reported on job labor cards. Attendance hours are usually reported as either Direct Labor Hours (actual job hours charged to a production order) or Expense Hours (time charged to expense or capital accounts and not against productive job hours.)

8. <u>Standard Hour (SH)</u> - a Standard Hour may be defined as the mutually agreed (management labor) time allowed to perform the elements of work at an incentive pace and including proper allowances for the conditions of work (crane wait, multiple machine or crew interference, tool crib, time reporting, etc.)

9. <u>Routing</u> - a planned order of performing defined operations for making a given product part.

10. <u>Standard Cost</u> (Industrial Engineered Standard Cost) - is the cost in dollars for the material and labor which whould be used on a job as defined by the Routing. Material dollars are the extension of the defined amount of material (pounds, inches, etc.) by the appropriate current material cost rate. Similarly, the Labor dollars are the extension of the rated (and unrated) hours defined on the routing by the appropriate labor and burden rates. Thus, the Standard Cost of a product can become the goal of planning of the corporate team or a cost reduction committee Thus, the effects of process changes can be measured in planning and, thus, become a tool for maximizing profitability.

11. **Load Hours (L)** - are the estimated hours (DWH) assigned to unrated operations on a routing for the purpose of arriving at a product standard cost. The term "load hour" was first used to define hours assigned to unrated operation for scheduling purposes and to estimate the work load allocated to operations on a specific shop facility.

TABLE OF COST SYMBOL DEFINITIONS

1. S = the total hours (rated standard hours and unrated day work) allowed for the performance of a specified job under defined conditions, and called <u>Standard Cost Hours</u>.

2. R = the reported hours (earned productive hours, EPH) on a job order.

3. E = the total earned productive hours reported on a job order.

4. C = the total clocked hours reported on a job order.

5. V = the numerical difference between the standard cost hours (S) and the reported earned productive hours (R) on a given job order. (See #10 below)

   $V = S - R$

   Special Condition -- when the Job Order includes parts previously made on Stock Orders, then. . .

   S = Total <u>Job</u> Standard Cost Hours are made up of Standard Cost Hours of Stocked Parts (Ss) plus the standard hours for Job labor (Sj).

   $S = Ss + Sj$

7. Ss = the standard cost hours allowed for Stocked Parts on a given <u>job</u> order.

8. Sj = the total standard cost hours allowed for the job portion of the order.

9. R = reported earned productive hours composed of standard hours of stocked parts (Ss) plus the earned productive hours for job labor (E).

   $R = Ss + E$

10. <u>Job variances (V)</u>:

    $V = S - R$

    $V = (Ss + Sj) - (Ss + E)$

    When the stocked part hours (Ss) on the Job Standard and reported job are equal (this is the normally controlled condition) then

    $V = Sj - E$

    Labor Hour Performance Measures:

11. <u>Labor Cost Performance</u>. (L$Pf)

    $L\$Pf = \dfrac{S}{R} = \dfrac{(SH + DWH)}{EPH}$ for any single order or

$$= \frac{Ss + Sj}{Ss + E} \quad \text{for a Stock related job order}$$

Use: To measure Cost ($) Performance for establishment of a historical average performance and under similar conditions the expected future performance.

12. **Job Labor Performance:** (JPf)

    Essentially the same as #11 except Ss (Stocked Standard Hours) are removed from the numerator and denominator.

    $$JPf = \frac{S}{R} \quad \text{(job orders w/o stock parts)}$$

    $$= \frac{Sj}{E} \quad \text{(orders with Ss Stock Parts labor removed)}$$

    Uses: Best comparative measure of effective labor cost performance, but (#11) may be more practical. . .because of the extra arithmetic operations required where stocked parts are involved.

    (#12) Puts stock related job orders on an equal basis with regular job orders since the Ss stocked part hours in numerator and denominator inflates the performance of stock related jobs.

13. **Job Productivity Index (P. I.)**

    (a) $\quad P. I. = \dfrac{S}{C} \quad$ (for stock orders w/o stock sub parts)

    (b) $\quad\quad\quad = \dfrac{Sj}{C} \quad$ (for all product sub orders)

    Use: Primarily to measure the expected continuous job labor performance improvement (Learning curve)

    The secondary purpose of recording (C) clocked job time is its suitability for scheduling job hours; e.g.,

14. **Job Scheduling Factor** -

    $$ScF = \frac{C}{S} \quad \text{for stock orders and product sub order or operations}$$

    $$= \frac{\text{Job clocked Hours (Meas. \& Unmeas. Direct Labor)}}{\text{Job (order) Standard Hours}}$$

Thus, when the long term average value of this factor is known, the expected clocked hours can be determined from the precosted standard.

To be effective, this measure would need expansion for stock scheduling.

EXAMPLE OF JOB PERFORMANCE

An example of reported labor hours on two successive job orders for the same product item will show how labor performance can vary:

Job hour standard
    Rated Incentive hours = SH   8.00 (SH)
    Estimated Day Work hours = DWH   2.00 (L)
    Total Job standard = S=SH+DWH   10.00 (S)

Job # 1

    Reported Earned Incentive hours = ESH   8.00 (E)
    Actual hours on Incentive (clocked)   5.00 (C)
    Reported Daywork hours = A(DWH) = C   3.00 (E) (C)
    Total Reported Earned Productive hours=EPH   11.00 (R)
    Total Clocked hours   8.00 (C)

R = ESH+DWH = 8+3 = 11
C = Actual Hrs. on Incentive + DWH = 5+3 = 8

Labor Performance (S/R) = (10/11)100   =   91%
Labor hours variance (S-R)   =   -1.00 (V)
Job Productivity Index (S/C)   =   125%

Job # 2

    Reported Earned Incentive hours   =   8.00 (E)
    Actual hours on Incentive (clocked) =   7.00 (C)
    Actual daywork hours reported   =   2.00 (E) (C)
    Total Reported Earned Productive hours =   10.00 (R)
    Total Clocked hours   =   9.00 (C)

Labor Performance (S/R) (10/10)   =   100%
Labor hours Variance (S-R)   =   0 (V)
Job Productivity Index (S/C)   =   111%

Development And Use Of Machinability Data For Process Planning Optimization

By Vijay A. Tipnis
Director of Manufacturing Technology
Metcut Research Associates, Inc.
and
Michael Field
President
Metcut Research Associates, Inc.
and
Moshe Friedman
Senior Lecturer
Technion, Haifa, Israel

During process planning for a computer aided manufacturing system, the selection of cutter paths, tools, workpiece material, cutting fluid, and operating conditions, such as speed, feed and depths of cut, should be made so as to minimize the total manufacturing cost. This requires a comprehensive analysis to identify the factors and variables that give rise to the major manufacturing costs. Machinability data is one of the important inputs needed to evaluate various alternatives for cost minimization. A recently introduced technique of establishing cutting rate-tool life characteristic curves allows rapid identification of the minimum cost region when a simultaneous choice of machining variables is required.

INTRODUCTION

The recent worldwide drive toward computer aided manufacturing (CAM) has drawn considerable attention to the development and use of machinability data for process planning optimization. Process planning, computer assisted or automatic, has been regarded as an important starting point for implementation of CAM. The objective of process planning optimization is to develop the most economical process plan, including optimized machining conditions, cutter paths, machining sequence, and tool replacement strategy, for manufacture of components to the desired dimensional tolerances, surface finish, and surface integrity.

One of the important relationships required for process planning optimization is the relationship between tool life and machining conditions. It is well recognized that at the current state of the machinability theory, the only reliable way to obtain this relationship is to conduct actual machining tests, either in the shop or laboratory. Based on the available machinability shop or laboratory data, a number of computerized machinability data systems have been introduced in the past.[1]* Some of these systems have been used to develop optimized machining conditions required for process planning optimization.[1,2] However, the usefulness of these systems has been limited due to 1) lack of reliable tool life data and 2) oversimplifying assumptions of tool life models such as Taylor relationships.

A comprehensive model of a computerized National Numerical Machining Data Bank for use in process planning optimization has been introduced.[3] A block diagram of the Machinability Data Bank is shown in Figure 1. The Machinability Data Bank generated optimized machining conditions required for process planning optimization by accepting input data from process sheets and from machinability, machine tool, tooling, and cost data files.

---

* Numbers in parentheses indicate References.

The Machinability Data Bank consists of three modules: Machinability Data File (MDF), Model Building (MB), and Optimization (O). The Machinability Data File consists of numerical data of machinability properties (tool life, surface finish, forces, deflections, etc.) for various machining conditions grouped according to workpiece material and machining operation. The Machinability Data File is continuously updated with fresh data from the 1) Machinability Data Center, 2) shop, 3) laboratory experiments, and 4) literature. The Model Building module consists of various routines for the development of mathematical models of tool life, surface finish, force, and other machining parameters. The Optimization module builds the target function (cost/piece, production rate, etc.) and imposes the constraints of surface finish, cutting forces, deflections, and horsepower, and then optimizes the target function within these constraints.

Until recently, most optimization attempts have used the classical machining economics approach.[1,2] This approach typically involves determination of cutting speed at which either the total cost is minimum or the production rate is maximum. The other cutting variables such as feed and depth of cut are held constant. For optimization involving multiple cutting variables, mathematical programming techniques are being applied.

Now, a more comprehensive approach is available for rapid identification of the minimum cost region where simultaneous choice of machining variables is required.[4,5,6] This approach uses second order linear (in logarithmic terms) tool life models and the newly developed concept of cutting rate-tool life (R-T) characteristic curves. This paper demonstrates how this new approach is used for developing and using machinability data for process planning optimization.

DEVELOPMENT OF TOOL LIFE MODELS

One of the important steps in determining machining costs and production times and in determining optimum machining conditions is to generate accurate and extensive machining data describing tool life as a function of machining conditions: speed, feed, depth, etc. Most investigators use Taylor's equation of the simplified form $VT^n = c$ or advanced form $VT^n F^\alpha D^\beta = c$:

where V = cutting speed (fpm)
T = tool life (min)
F = feed (ipr or ipt)
D = depth of cut (in)
c = a constant for a given combination of workpiece and cutting tool
n, α and β are the coefficients

It has been demonstrated that the Taylor's equation does not fit the actual tool life data for many important commercial alloys when the range of cutting speed, feed, and depth of cut is wide enough to be of practical value.

The reason for the lack of fit is that the tool wear is produced by different wear mechanisms at different combinations of speed, feed, and depth of cut.

A more comprehensive tool life relationship (given below) has been found to fit most tool life data of many commercial alloys over the wide range of speed, feed, and depth of cut often found in practice. This tool life relationship has the following general form of second order linear (in logarithmic terms) equation:

$$\begin{aligned}
\ln T = {} & b_0 + b_1 \ln V + b_2 \ln F + b_3 \ln Da + b_4 \ln Dr \\
& + b_{11}(\ln V)^2 + b_{22}(\ln F)^2 + b_{33}(\ln Da)^2 + b_{44}(\ln Dr)^2 \\
& + b_{12}(\ln V)(\ln F) + b_{13}(\ln V)(\ln Da) + b_{14}(\ln V)(\ln Dr) \\
& + b_{23}(\ln F)(\ln Da) + b_{24}(\ln F)(\ln Dr) + b_{34}(\ln Da)(\ln Dr)
\end{aligned}$$

-----(1)

where T = tool life (min)
 V = speed (fpm)
 F = feed (ipt)
 Da = axial depth (in)
 Dr = radial depth (in)
 $b_0, b_1, \ldots\ldots\ldots, b_{34}$ are the coefficients

The development of a mathematical tool life model using Equation (1) involves conducting statistically planned tool life tests and curve fitting by stepwise multiple regression analysis.[7] It has been demonstrated that far fewer tool life tests are necessary when a well defined statistical plan is used instead of the classical plan of conducting tests by studying only one variable at a time. Also, this stepwise regression analysis evaluates and accepts only the most influential of the terms given in Equation (1), thus building a compact yet comprehensive tool life model. A number of such comprehensive tool life models have been developed.[8] The development of machining data such as tool life data using these comprehensive mathematical models is an essential task involved in building a computerized Machinability Data Bank that can be used for process planning optimization.

THE CONCEPT OF R-T CHARACTERISTIC CURVES

The concept of cutting rate-tool life (R-T) characteristic curves emerged when the lines of constant tool life were plotted in the plane of speed (V) versus feed (F).[8] These lines of constant tool life were obtained from a second order tool life model. When the lines of constant cutting rate, i.e., the material removal rate (say, in cu. in. /min.) for that machining operation, were superimposed on the V-F plane, it became evident that there are regions in the V-F plane where better combinations of cutting rate and tool life can be found.

The concept is best illustrated through a plot such as shown in Figure 2, where the constant cutting rate (R) and constant tool life (T) lines are drawn.

At the points where constant R lines are tangents to constant T lines, the concept states that for any given cutting rate, tool life is maximum, and for any tool life, the cutting rate is maximum. The line joining the tangent points is called the cutting rate-tool life characteristic curve. It has been proven that the economic optima, such as minimum cost, maximum production rate or maximum profit, must lie on the R-T characteristic curve.[4] This is illustrated in Figure 3 where the R-T characteristic curve and the position of optimum cost and maximum production rate points are shown.

The concept of the R-T characteristic curve leads to numerous potentially useful, practical applications,[5] most important of which are 1) the rapid identification of economic machining regions and 2) the introduction of a comprehensive method for economic optimization when simultaneous choice of numerous machining variables is involved. A detailed treatment of R-T characteristic curves, their verification and application, is found elsewhere.[4, 5]

DETERMINATION OF COST AND PRODUCTION RATE IN MACHINING

The classical machining economics involves consideration of cost, production rate or profit rate as a function of all machining variables such as speed; feed; depth, width, length of cut; labor and overhead rates; and tooling and tool reconditioning costs. Comprehensive cost and production time (or production rate) equations for several conventional and N/C machining operations have been developed[9] and have been incorporated in a computer program called NCECO. These comprehensive cost and production time equations and the NCECO program give a detailed breakdown of ten costs associated with each machining operation. It has been demonstrated that these equations and the program can be used to 1) identify the most significant items where cost and time expenditures are high and 2) select speeds, feeds, cutting tools, and work materials that give lower costs and higher production rates.

The application of R-T characteristic curves for identification and optimization of economic machining conditions uses the same cost and production rate equations as in the classical approach but in a modified generalized form given below:

$$\frac{1}{P} = m_0 + \frac{m_1}{R} + \frac{m_2}{RT} \quad \text{-----(2)}$$

$$C = k_0 + \frac{k_1}{R} + \frac{k_2}{RT} \quad \text{-----(3)}$$

where C = cost/piece
       P = production rate/time

R = cutting rate (cu in/min)
T = tool life (min)
m_0 = idle, load-unload, and setup time per piece
m_1 = volume of workpiece material machined + volume due to extra travel at feed
m_2 = volume of workpiece material machined (down time to replace worn tool)
k_0 = m_0 (labor + overhead per unit of time)
k_1 = m_1 (labor + overhead per unit of time)
k_2 = m_2 (labor + overhead/time) + (volume cut)(tool and cutter reconditioning cost)

In these equations: $R = f_1(V, F, Dr, Da)$
$T = f_2(V, F, Dr, Da)$

The existence of the R-T characteristic curve implies that $T = f(R)$. A formal proof and derivation of $T = f(R)$ and optimization of cost and production rate functions given in Equations (2) and (3) are given elsewhere.[4] To illustrate the use of R-T characteristic curves, the following example involves 1) experimental determination of second order tool life equation, 2) determination of R-T characteristic curve, and 3) cost and production rate functions.

IDENTIFICATION OF ECONOMIC MACHINING REGION FOR PERIPHERAL END MILLING OPERATION

Consider peripheral end milling of an annealed 4340 steel using a standard M2 HSS 1 in. diameter, 2 in. flute length, 4 flute end mill. To determine a second order tool life relationship, a series of statistically planned experiments in the region of speed, V = 50 to 150 ft./min., feed, F = 0.004 to 0.010 ipt, radial depth, Dr = 0.1 to 1.0, and axial depth, Da = 0.5 to 1.5, were performed on a Cinova 80 vertical milling machine. A water soluble fluid was used throughout the tests. Several tests were performed at different combinations of speed, feed, radial depth, and axial depth according to a statistical experimental design. Using the stepwise regression programs previously described,[7] the following second order tool life model was obtained:

$$(\ln T) = -48.9 + 21.59(\ln V) + 21.50(\ln Da) + 3.40(\ln F)^2$$
$$+ 2.33(\ln V)^2 + 7.99(\ln V)(\ln F) + 4.03(\ln F)(\ln Dr)$$
$$+ 4.55(\ln F)(\ln Da) + 3.87(\ln V)(\ln Dr) \quad \text{-----(4)}$$

The residual standard deviation for the fit with experimental data was 0.24 which is reasonable considering that the variance of tool life data for this type of operation is about 20 percent. It should be noted that this second order equation is valid for tool life values of 15 to 60 minutes with the range of speed, feed, radial depth, and axial depth used in the tool life tests. Although interpolation is permitted within this range, extrapolation beyond this range is not considered valid.

The cutting rate in peripheral end milling is given by:

$$R = \frac{12}{\pi} \frac{FVt}{d} DrDa \quad\quad\quad -----(5)$$

where t = number of flutes
d = diameter of the end mill

R, F, V, Dr, and Da have been defined before.

The lines of constant tool life (bold lines) and the lines of constant cutting rate (dotted lines), as obtained from Equations (4) and (5), respectively, are plotted in Figure 4 in the V-F plane at Dr = 0.30 in. and Da = 1.25 in. The R-T characteristic curve for the operation was plotted using analytical and iterative methods described elsewhere.[5] It can be seen from Figure 4 that the R-T characteristic curve lies in the practical range of speed, V = 70 to 90 ft./min., and feed, F = 0.004 to 0.007 ipt, for this operation. Along the R-T characteristic curve, the tool life is maximum for any given cutting rate and vice versa.

It should be noted that the region where tool life tests were conducted was established and revised through a series of intermediate tool life models and R-T characteristic curves before the final tool life model given in Equation (4) was obtained. Most tool life tests were conducted in the range of 15 to 80 minutes tool life. As can be seen from Figure 4, in the practical range of speed and feed, additional tests in the vicinity of the R-T characteristic curve should help define the economic region more precisely. Thus, the R-T characteristic curve enables identification of the ranges of speed and feed where economically optimum conditions are likely to occur.

AN EXAMPLE OF DETERMINING OPTIMIZED MACHINING CONDITIONS

To illustrate the use of the R-T characteristic curve for determining optimized machining conditions, consider an annealed 4340 steel bracket shown in Figure 5. This part is end milled using a 1 in. diameter, 2 in. flute length, 4 flute M2 HSS end mill. Water soluble cutting fluid is used during the end milling operation. To simplify, consider only three passes of the end mill at the total length of about 140 in. around the part. The radial and axial depths of each pass are .030 in. and 1.25 in. respectively. Obviously, the part is machined through several roughing and finishing passes and is subjected to other operations such as drilling, face milling, etc. For this example, however, we shall confine our attention to the three passes stated above. We shall also omit all setup and load-unload operations from our discussion, since their times and costs are independent of cutting rate and tool life.

The tool life model given in Equation (4) and the cutting rate in Equation (5) were used for determining cost and production rate for the three passes. The time and cost data used for determining the coefficients m_0, m_1 and m_2, and k_0, k_1 and k_2 of the production rate and cost Equations (2) and (3), respectively, were as follows:

- Labor + overhead at N/C milling machine = \$0.83/min.
- Labor + overhead at cutter sharpening department = \$0.17/min.
- Time to replace dull cutter = 20 min.
- Cutter cost = \$56.00
- Number of resharpenings = 4
- Time to resharpen = 15 min.
- Extra travel in feed = 5.5 in.
- Rapid traverse rate = 150 in./min.

Each cost and production time item was determined using the NCECO program.[9] The coefficients of Equations (2) and (3) obtained are given below:

$$m_0 = 0.18 \qquad k_0 = 0.15$$
$$m_1 = 54.37 \qquad k_1 = 45.13$$
$$m_2 = 1050. \qquad k_2 = 1598.$$

The results of the computer analysis are plotted in Figures 6, 7, and 8. In Figure 6, the lines of constant cost for the three passes (bold lines) and the lines of constant production rate (dotted lines) are shown in the economic region of speed and feed. Note that the cost curves are closed ellipses with the midpoint on the R-T characteristic curve. Note that just as for any given cutting rate, the tool life is longest on the R-T characteristic curve; for any given production rate, the cost is lowest on the R-T characteristic curve. Furthermore, the speed and feed coordinates of the minimum cost point on the R-T curve give the optimized milling conditions.

In Figure 7, the costs at different combinations of speeds and feeds along constant cutting rate curves and along the R-T characteristic curve are shown. It can be observed that the cost decreases as the R-T characteristic curve is approached from either side along the constant cutting curves. The cost values along the R-T characteristic curve pinpoint the location of the minimum cost.

The same cost data is plotted against cutting rate in Figure 8 to illustrate the cost minimum.

It should be noted that in this example all the cost and production rate values are given in a fairly well identified economic region. At speeds and feeds other than in this region, the costs are much higher.

The entire procedure, after experimentally establishing the second order tool life model, was carried out through several computer routines that constitute a major part of the optimization model currently being developed. Referring back to Figure 1, the task of determining optimized machining conditions is accomplished comprehensively and rigorously by using R-T characteristic curves.

Acknowledgment: The assistance of Mr. Steve Buescher in computer programming and of Messrs. John Kohls and Steve Buescher in preparing illustrations is gratefully acknowledged.

REFERENCES

(1) Weller, E.J., Ritz, C., Montaudouin, J., Hirsch, B.E., Zölzer, H., Engelskirchen, W.H., and Zimmers, E.W. 1971. Computerized machinability data systems. In N/C Machinability Data Systems, ed. N.R. Parsons, pp. 63-128. Dearborn, Michigan: SME.

(2) Barash, M.M. and Batra, J.L. 1972. Automatic optimal process planning--the next step in CAM. SME Paper No. M572-169.

(3) Friedman, M.Y., Field, M., and Kahles, J.F. 1974. Machinability data bank design. Annals of CIRP 23:171-172.

(4) Friedman, M.Y. and Tipnis, V.A. 1974. Cutting rate-tool life characteristic functions for material removal processes: Part I - theory. Submitted for publication in ASME Transactions.

(5) Tipnis, V.A. and Friedman, M.Y. 1974. Cutting rate-tool life characteristic functions for material removal processes: Part II - verification and applications. Submitted for publication in ASME Transactions.

(6) Tipnis, V.A. and Friedman, M.Y. 1975. Cutting rate-tool life characteristic functions for three machining variables--theory and applications. Paper to be presented at NAMRC-III, Pittsburgh, May 1975.

(7) Friedman, M.Y. and Field, M. 1974. Building of tool life models for use in a computerized numerical machining data bank. In Proceedings of the International Conference on Production Engineering, Tokyo 1974 (Part 1), pp. 596-601. Tokyo: Japan Society of Precision Engineering.

(8) Tipnis, V.A., Christopher, J.D., and Friedman, M.Y. 1974. Experimental investigation of tool life of circular saws: Part I - effect of speed and feed using statistical methods. In Proceedings of NAMRC-II, pp. 154-164. Dearborn, Michigan: SME.

(9) Ackenhausen, A.F. and Field, M. 1970. Determination and analysis of costs in N/C and conventional machining. SME Paper No. MR70-545.

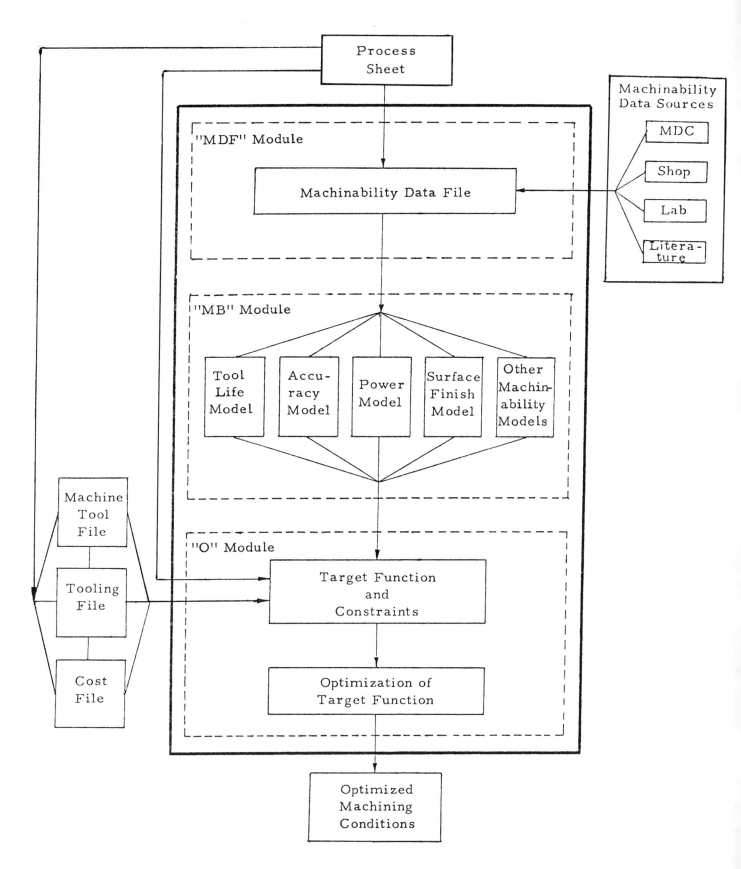

Figure 1. Block Diagram of Computerized National Numerical Machining Data Bank (from Ref. 3)

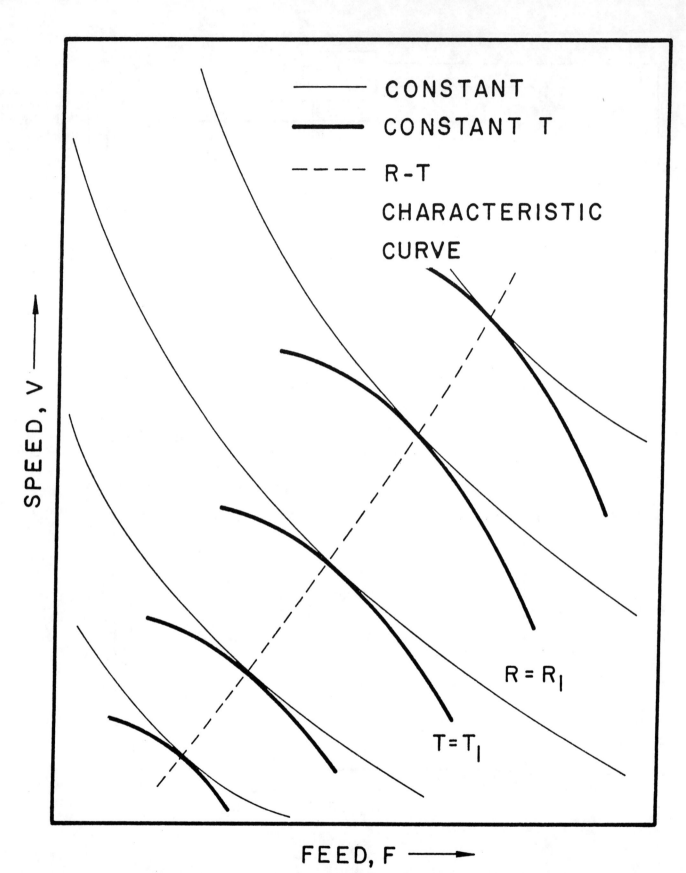

Figure 2. R-T Characteristic Curve in V-F Plane

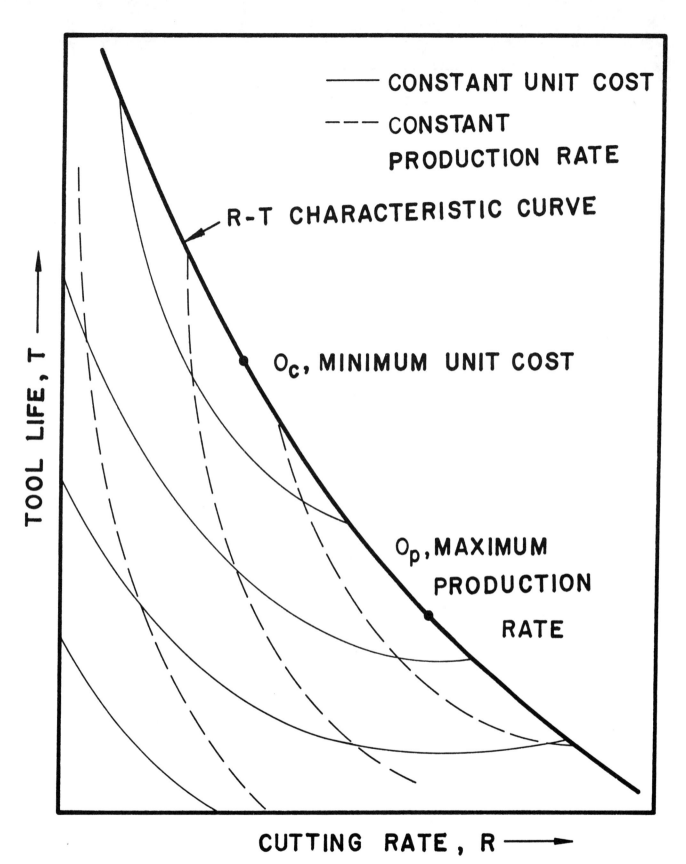

Figure 3. R-T Characteristic Curve in R-T Plane Showing Minimum Cost and Maximum Production Points

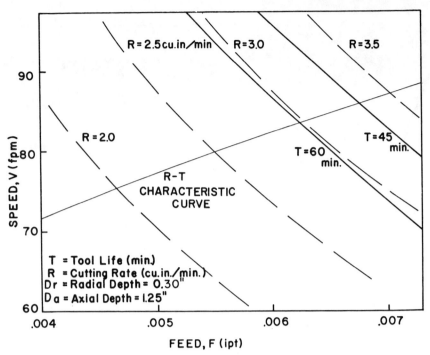

Figure 4. R-T Characteristic Curve for End Milling an Annealed 4340 Steel

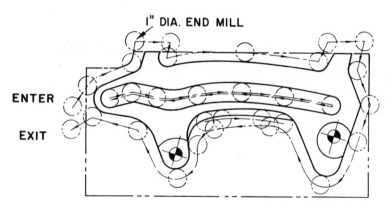

Figure 5. Bracket Machined from a 4340 Annealed Steel Blank using Peripheral End Milling Cuts with 1 In. Diameter, 2 in. Flute Length, 4 Flute M2 HSS End Mills

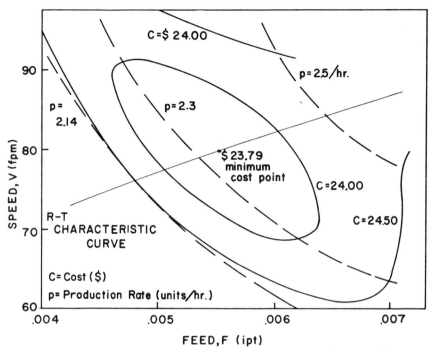

Figure 6. Cost and Production Rate Profiles on V-F Plane with R-T Characteristic Curve

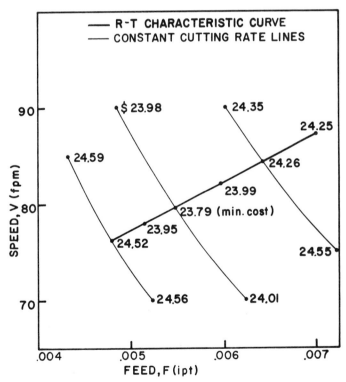

Figure 7. Cost Along Constant Cutting Rate Curves in V-F Plane

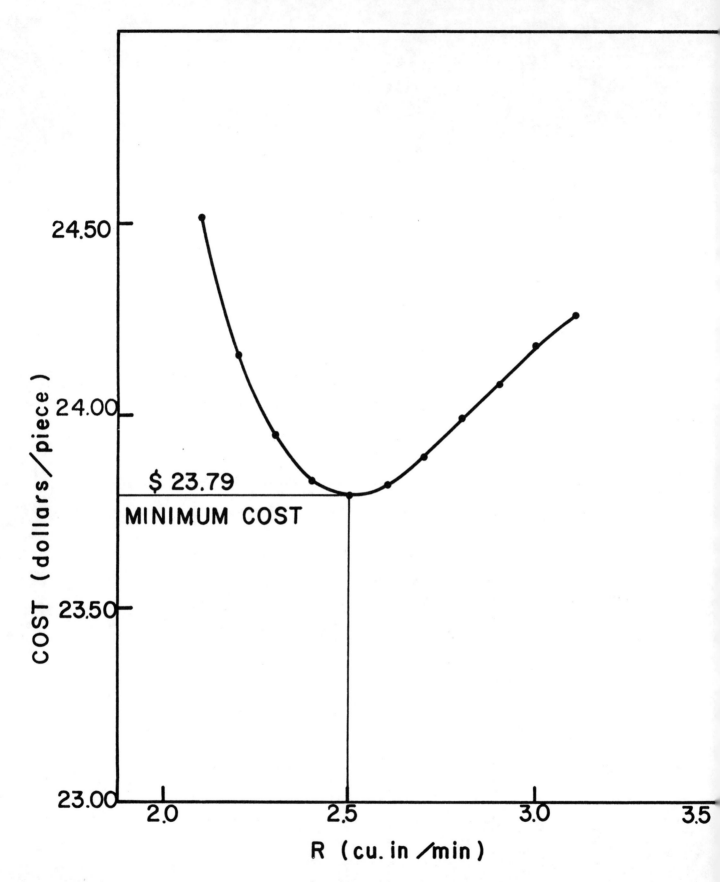

Figure 8. Cost Versus Cutting Rate

Phillip F. Ostwald, Ph.D., Member AACE
Professor

Patrick J. Toole, Ph.D.,
Member AACE
Assistant Professor

Department of Engineering Design and Economic Evaluation
University of Colorado at Boulder
Boulder, Colorado

A PRIMER ON MATERIAL ESTIMATING

INTRODUCTION

This paper summarizes nine methods of assigning a cost to a particular material used in the job estimated. The complexities of this problem are first reviewed. Distinctions between manufacturing and contracting estimates as they impact the selection of a material cost are considered. The methods are previewed, and then a specific problem having costs, requirements, and inventory levels is formulated. Each of the methods is discussed within the contect of this problem. The last method presented is new and widespread consideration by industry is encouraged.

IMPORTANCE OF PROBLEM

Fundamentally, the problem of estimating materials is to separate the problem into two steps, i.e., find the shape requirements and multiply that by the cost per unit. Shape or mass, or area, or length, etc., is expressed in various engineering dimensions. There is variety in the ways to calculate shape along with an allowance for scrap, waste, and shrinkage, but the focus of this paper is on the second step: that of finding the cost for materials.

Uncertainty in price has been a factor of increased importance since 1972 due to higher rates of inflation. Furthermore, some materials are unavailable and shortages are occurring more frequently. Some companies have encountered financial losses after seeing their raw material costs escalate rapidly.

The material costing problem is important in other aspects. When considering lists of materials, finishes, standard and non-standard designs, and recalling the inventories of direct and indirect materials, it is not surprising to find companies that have an active material list of many thousands. Inventories are extensive, thus complicating the finding of costs for materials in an estimate.

Many estimating functions have been unable to cope with the problem of picking a cost value for a material. Cost values using accounting systems, engineering code classifications, vendors, government indexes, and shared professional, consulting or association information can add to the bewilderment of picking a cost value. The purpose of this paper is to shed some light on the problem of providing a taxonomy of methods for the materials estimate problem, i.e., what is the cost per unit for the material?

DISTINCTIONS BETWEEN MANUFACTURING AND CONTRACTING ESTIMATES

Manufacturing materials often have different technical specifications than construction materials (e.g., SAE 1035 vs A36). Furthermore, a manufacturing firm may build a product from a continuously replenished inventory, while a contracting firm may not order materials until the bid has been won. For cost estimation purposes, the differences in classifications and in material ordering policies are to some extent superficial. For example, a job shop firm, or a contractor, will often not order many of its materials until a particular job is won, and a manufacturing firm may special order materials in building prototypes even though the equipment may be unsold at the time. The methods that follow are intended to suit the needs of both manufacturing and contracting. The word "lot" as used subsequently is more commonplace in manufacturing than in contracting. It implies a purchase order of a quantity of discrete items, perhaps as few as one.

A SPECIFIC EXAMPLE

The problem to be considered in illustrating the methods for material estimating is given in Figure 1. The horizontal axis is time. The time units may represent days, weeks, years, etc. The present is to be time 0 and is labeled "E" to designate it as the time of the estimate. Negative integers are past time epochs and positive integers are future time epochs. Point D is the delivery time for the product. The vertical axis represents units of quantity of the material.

The three functions designated by TS, TU and JJ are defined as follows:

1. TS(t) = total supply of the material at time t.
2. TU(t) = total usage of the material at time t.
3. JJ(t) = job J usage of the material at time t.

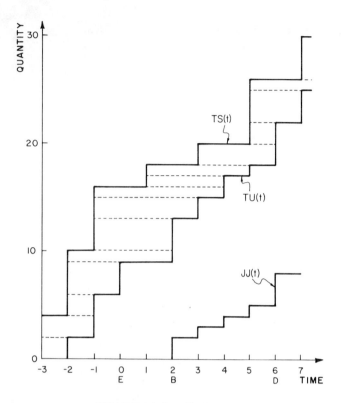

FIGURE 1 A Specific Example

These step functions will be used to obtain different estimates for the unit material cost for Job J.

To the right of point 0, each of the functions represent estimates. The step sizes on each of the integers t = 0, ±1, ±2, ..., for the functions TU and JJ represent the material used in production during the period (t-1, t). The step sizes on each of the integers t for the function TS represent the quantity of inventory arriving at time t. Hence the inventory at time t = 0, ±1, ±2, ..., is the difference between the functions TS and TU during the period (t, t+1).

In this example it is assumed that the total usage curve, TU, is comprised of the job J curve, JJ, plus other curves. For example, at t = 4, job J uses 50 percent of the total for that period, and at t = 5, job J uses 100 percent. Material estimates are made assuming job J is won because, as will be expanded on in the next section, quantity discounts are assumed to exist for material purchases.

The inventory supply curve TS does not follow a pre-established review and reorder discipline such as a reorder point system or a periodic review system (see Reference 3, Chapter 4) but rather, is assumed as customarily found in estimating to depend on factors external to the information presented. A fixed lead time of one period is assumed for all orders.

PROJECTING UNIT COSTS

Table 1 displays realized and projected unit costs the firm faces in purchasing the materoal. Price breaks are available for various quantity lots. Two methods will be presented for projecting unit costs in each of the price break categories. The second of these methods is the one used to derive the projected unit costs of Table 1.

	Period	Quantity			
		1 - 2	3 - 4	5 - 6	7 - 8
Realized	-4	10.00	8.00	7.00	6.50
	-3	10.70	8.60	7.50	7.00
	-2	11.10	8.95	7.80	7.20
	-1	11.25	9.00	7.90	7.30
	0	12.00	9.65	8.40	7.85
Projected	1	12.95	10.01	8.73	8.12
	2	12.97	10.44	9.10	8.46
	3	13.52	10.89	9.49	8.83
	4	14.10	11.36	9.89	9.20
	5	14.69	11.84	10.32	9.60
	6	15.31	12.35	10.75	10.01

TABLE 1: Realized and Projected Unit Costs

The moving average projection method establishes, as an estimate of the future cost for one or more periods, the average of the present plus a number of immediately preceding period costs. After a period has elapsed, a new average is calculated by dropping the oldest cost and adding in the newest. For example, at t = -2, a three period moving average for the 1-2 purchase range would be

$$(10.00 + 10.70 + 11.10)/3 = 1060,$$

and at time 0 the same average would be
(11.10 + 11.25 + 12.00
$$(11.10 + 11.25 + 12.00)/3 = 11.45.$$

Moving averages are useful when the data does not show any forecastable trend.

The linear regression trend is used when a linear trend is apparent from the data (for a considerably more sophisticated analysis, see Reference 1). After plotting the data in

Table 1 for each of the price break categories for t = -4, -3, -2, -1, 0, the following linear model is established

ln P9t) = a + b (t-4)

where P(t) is the uni material price projection at time t. This relationship is reasonable assuming a constant rate of inflation 1 + x, because in that case

P(t) = P(-4) (1 + x)$^{t-4}$.

Using standard statistical techniques we estimate a and b using the five years of available data. The resulting projected figures are given in Table 1.

A PREVIEW OF THE METHODS

The nine methods to be used for selecting a material cost may be categorized into three groups:
1. Point-in-Time Costs.
    a. Original.
    b. Last.
    c. Current.
    d. Lead Time Replacement.
    e. Delivery.

2. Contractual Costs.
    a. Quotation.
    b. Quote or Price-in-Effect.

3. Realized Costs.
    a. Money-out-of-Pocket.
    b. True Inventory Cost.

POLICIES TO PICK COST OF MATERIALS

Some of these policies were developed by the accounting profession for the purpose of inventory valuation necessitated by tax and ownership laws. However, accounting practices have different goals than estimating and caution must be exercised. Furthermore, the many accounting systems, often called standard cost, job or lot or process costs, are procedures and are not ploicies to determine a future cost for materials. The methods to follow do include as special cases many standard industrial cost procedures which emphasize a predetermined value for the material. The methods involved in job, lot, or process cost systems are not included because their basic function is to assess the material cost after production, which is not the basic estimating problem.

ORIGINAL COST

This method assumes that materials are used in the order they are received and establishes as a cost estimate the unit cost of the oldest material in inventory. In Figure 1 the oldest material in stock at time E cane from a lot of six purchased at t = -2. From Table 1 the unit cost is $7.80. Commonly called FIFO, this method has been popularized by the accounting profession for inventory valuation purposes.

LAST COST

The Last Cost method assumes that the latest materials purchased are the first to be used and establishes as a cost estimate the unit cost of the most recent material in inventory. In Figure 1 the most recent material in stock at time E came from a lot of six purchased at t = -1. From Table 1 the unit cost is $7.90. This method, called LIFO, is frequently used by the accounting profession.

CURRENT COST

A unit cost at time 0 is used in this method as the value of the estimate and thus is time coincident to the establishment of the estimate. If price breaks exist, a lot size must be determined before an estimate is arrived at. For example, if the purchase lot is assumed to be 6 at time 0 (even though no material purchases are made in Figure 1) the estimate from Table 1 is $8.40/unit.

LEAD TIME REPLACEMENT COST

This concept adopts as an estimate the replacement cost of the first lot of the material at the time it arrives. This is commonly called the next-in-first-out method (NIFO) even though this label is obviously an anomaly. Suppose in Figure 1 that the estimate is won and a lot of four is ordered at t = 1. Since there is a lead time requirement of one period, the order arrived at t = 2 and the unit cost estimate from Table 1 is $10.44. This concept is used by product estimators where material renewals are significant.

DELIVERY COST

The point in time at which the production order is delivered to the customer establishes the value for the estimate. As in the above methods, a lot size must be assumed if price breaks in material purchasing are significant. In Figure 1 the delivery time is point D and if we assume a lot size of four the unit cost is taken as $12.35.

QUOTATION COST

The Quotation Cost method is widely practiced by job shops and contractors. The cost of the material is established by a delivery time quotation from a vendor. This method is similar to the lead time replacement cost given above, except that the vendor of the material makes the forecasts as those given in Table 1, the unit price would be $10.44. The delivery price quoted is usually considered firm, unless contractual provisions exist for quote or price-in-effect adjustments.

QUOTE OR PRICE-IN-EFFECT

Known for many years, this concept has become increasingly popular during recent years of high inflation and rapidly escalating material prices. A collaborative legal agreement between the buyer and seller is usually established which allows for adjustments to the original price should the seller incur material costs in excess of those estimated. If any material costs fall below their estimate, the buyer agrees to the original price and adjustment is unnecessary. Any of the other methods in this paper may be used to establish the original estimate. If at the time of delivery of the product the seller can prove that the material costs incurred were in excess of the original estimated value, the buyer will make up the difference via formula or full cost allowances. The quote or price-in-effect method may avoid the often difficult problem of making accurate forecasts, however, competition and bidding may not allow the luxury of this policy.

MONEY-OUT-OF-POCKET COST

Money-out-of-pocket refers to a general class of techniques in which the overriding philosophy is to have the estimate reflect actual material expenditure. If the material is purchased in a single lot then the estimate would simply be the unit cost of the lot. If the material is taken from inventory then the estimate would be the original purchase cost of the material used. If the material is to be purchased in the future, a forecast of the unit price to be realized would be used.

TRUE INVENTORY COST

A new method that is actually a specific implementation of the out-of-pocket concept is the True Inventory Cost method for materials inventory. In the example of Figure 1, the material for job J is actually taken from a number of past and future inventory purchases. The True Inventory Cost method accounts for forecasted usage of the material in as accurate a manner as possible. Suppose at time t we have $n(t)$ lost of the material in inventory, which were purchased at times t_j; $j = 1, 2, \ldots, n(t)$. At time t define

$f(t_j)$ = the fraction of the material inventory coming from the time t_j lot

$p(t_j)$ = unit cost of the t_j lot

$d_J(t)$ = number of units of the material used for job J at time t.

The estimated unit materials cost is:

$$\sum_t \sum_{j=1}^{n(t)} p(t_j) f(t_j) d_J(t) / \sum_t d_J(t)$$

In the example, this would be:

$(7.80 \times 1/4 \times 2 + 7.90 \times 3/4 \times 2 + 7.90 \times 1 \times 1 + 7.90 \times 1/2 \times 1 + 10.01 \times 1/2 \times 1 + 10.01 \times 1 + 10.89 \times 1/2 \times 3 + 10.75 \times 1/2 \times 3) \div 8 = \$9.38$

Although this method requires careful inventory accounting and more computation, it yields the most accurate forecast in complicated and high risk situations. For most firms the additional inventory accounting and computation for material inventories of significant value would not pose a significant additional burden on the data processing system in use.

SUMMARY

The following table summarizes the results of applying each of the nine methods to the example of Figure 1.

METHOD	VALUE ADAPTED FOR ESTIMATE (in $)
Original	7.80
Last	7.90
Current	8.40
Lead Time Replacement	10.49
Delivery	12.35
Quotation	10.44
Quote or Price-in-Effect	*
Money-out-of-Pocket	*
True Cost	9.38

*Cannot be established without further assumptions.

Table 2: Unit Cost Estimates

REFERENCES

1. Draper, N. R. and Smith, H., Applied Regression Analysis, Wiley, 1966.

2. Ostwald, P. F., Cost Estimating for Engineering and Management, Prentice-Hall, 1974.

3. Schmidt, O. W. and Taylor, R. E., Simulation and Analysis of Industrial Systems, Irwin, 1970.

Four-slides, Part 6 — Estimating Your Costs

Estimating stamping costs for four-slide production operations can be done with pinpoint precision, using formulas discussed in this article. Importantly, these formulas can also be used to estimate the cost of production when stamping presses are used

ALL MANUFACTURERS of metal stampings have the same basic objective: to make required parts to users' specifications at the lowest possible unit cost. The ability to produce the part within allowable dimensional tolerances and other user specifications depends on properly designed tools and making the parts in the optimum type of equipment. Predicting the actual cost of making the parts — and therefore establishing an acceptable selling price — may be quite difficult and yet may make the difference between profit and loss in producing a particular order.

The production cost for producing any metal part includes five major components: (1) overhead, (2) labor, (3) equipment, (4) tooling, and (5) materials. Importantly, the costs of these components can be individually calculated, then combined to provide an accurate price quotation.

Overhead and Labor Costs. Overhead varies widely from company to company and also from time to time within any one company. Yet it covers many indirect costs which nevertheless must be considered when determining costs for quotes on metal stampings. While accounting procedures differ widely between companies and while they also are influenced by many factors, we can benefit from the approach suggested by E. H. Lanke of Wisconsin Coil Spring in a recent issue of *Springs* Magazine.

Mr. Lanke's approach is to assign indirect costs either to what he terms an "employee gross hourly cost" (labor cost) or to a "machine hour rate" (equipment cost). The employee gross hourly cost includes direct hourly wages, the costs of overtime, paid vacations and holidays, state and federal unemployment taxes, pensions and insurance. The machine hour rate includes machine value, rent, heat and light, general maintenance, support equipment, inspection area and equipment, and toolroom area and equipment. Additionally, a projected percentage profit is applied to both the labor and machine rates for estimating purposes.

Equipment Costs. The largest factor in equipment cost is machine value, the assigned hourly rate based on projected utilization over the selected depreciation period. If an engineer decides to purchase a machine, he is often tempted to invest in used equipment that costs one half to one third the price of a new machine. However, he must consider the additional cost of bringing the used machine up to good working condition. This includes the cost of machine production hours and the time that skilled personnel would otherwise be directly engaged in the production of stampings.

There may be a choice of several different models or types of equipment for making particular parts. In making his choice, the manufacturing engineer may have to consider whether or not his company should invest in a machine for its immediate advantages or for the greater long term advantages. For example, a limited capacity machine that can make a particular stamping may cost far less than a machine that, with the necessary tools and attachments, can be used to make a wide variety of other parts as well as the required stamping. By providing faster setup and higher production rates, the higher-cost machine may return its additional cost in a matter of months.

The manufacturing engineer may also be comparing the cost of buying a machine with the cost of an available machine already in the shop. Depending on relative variable and fixed costs, it may not be as economical to use the available machine. A break-even analysis to be described may reveal that it would be more profitable for a company to invest in a new machine to make one or more parts.

Tooling Costs. This variable is affected by the nature of the part, type of equipment, and the type and design of the tools themselves. Class *A* tools, which are made from high-quality tool steel with all sections hardened and ground, offer tool life in the order of 20 million pieces before major rebuild and replacement are required. Class *B* tools cost less but have shorter life expectancy, hence are practical only for short production runs. Most machine builders recommend that their equipment be used with Class *A* tools only.

Material Costs. Materials are the primary factor in costing (and pricing) stamped parts. The selection of a type of material that is not right for the part can add a hundred percent or more to the minimum part cost. For example, stainless steel can be expected to be three or four times as costly as carbon steel, which might possibly be surface finished to meet corrosion-resistance specifications. Depending on the skill of the designer, and type of equipment used, a large percentage of the material may be wasted. This, of course, adds to the cost of the stamping without contributing to its value.

Total Cost Formula. The formula for total cost per thousand stampings can be expressed by the following equation:

$$C = C_L + C_E + C_T + C_M$$

in which C = total cost per thousand parts, C_L = labor cost per thousand parts, C_E = equipment costs per thousand parts, C_T = tooling cost per thousand parts, and C_M = material costs per thousand parts. To use this formula, it is necessary to calculate the value of labor, equipment, tooling and material costs by individual formulas.

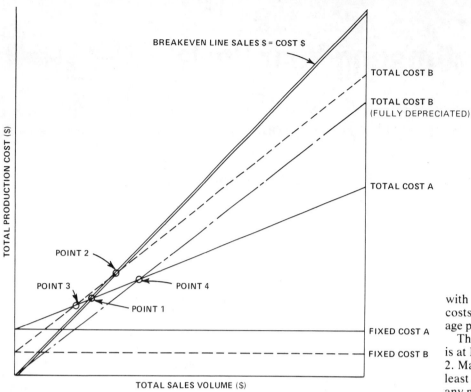

▶ LABOR COSTS. In calculating labor costs (C_L) the following formula is used:

$$C_L = \left(\frac{C_{LO}}{R} + \frac{C_{LS} \cdot T_S}{N_R}\right) \times 1000$$

In this equation, C_{LO} = operator labor cost (including applicable overhead) in dollars per hour. R = the net production rate in parts per hour. C_{LS} = setup man labor cost (including applicable overhead) in dollars per hour. T_S = time required for setup in hours. N_R = the number of parts in the production run.

▶ EQUIPMENT COSTS. Next, the equipment costs (C_E) are calculated. The formula used is

$$C_E = \frac{1000}{R}\left(C_{EO} + \frac{C_{ED}}{T_A}\right)$$

In this equation, R = the net production rate in parts per hours. C_{EO} = the machine overhead charge in dollars per hour. C_{ED} = the annual machine write-off for depreciation in dollars. T_A = the total annual production time in hours.

▶ TOOLING COSTS. The formula for calculating tooling costs (C_T) is as follows:

$$C_T = \frac{C_{IT}}{N_P} \times 1000$$

In this equation, C_{IT} = the initial tooling cost in dollars. N_P = the number of parts produced over the life of the tooling.

▶ MATERIAL COSTS. The final variable, material costs, (C_M), are calculated with the formula,

$$C_M = C_P \cdot D_M LWt \times 1000$$

In this formula, C_P = price of the workpiece material in dollars per pound. D_M = density of the material in pounds per cubic inch. L = feed length in inches. W = stock width in inches. And t = stock thickness in inches.

And thus the value of total cost (C) is easily arrived at by simply adding the values of C_L, C_E, C_T and C_M.

Break-even Analysis. Evaluation of two alternative methods of part production may be carried out in terms of potential sales volume from the machines rather than in relative cost per thousand parts. And as was previously mentioned, the fact that a machine costs less initially or happens to be fully depreciated may have little bearing on its real return on investment.

Determining which is the more profitable machine can be helped by a break-even analysis as is demonstrated for hypothetical machines A and B in the accompanying graph. The total production cost on the vertical scale includes both fixed and variable costs for the two machines. Fixed cost is the sum of depreciation, insurance, taxes and other overhead charges which do not change with machine output. Total variable cost includes labor and machine time for setup and running, maintenance and other factors which change with machine output. (All production costs are assumed to include a percentage profit margin.)

The breakeven point for Machine A is at Point 1 and for Machine B at Point 2. Management is losing money — or at least not meeting its profit goals — on any machine operating at sales volumes below the breakeven line (where production cost exactly balances out sales volume), and making money at sales volumes above the line, or beyond the breakeven point.

The higher productivity of Machine A, which results in a lower variable cost for any particular sales volume, was purchased at a higher initial cost and therefore has a higher fixed cost than Machine B. However, the total of the fixed and variable costs of Machine A decreases in relation to the total cost of the less efficient Machine B as sales increase. At the sales volume where the two total cost curves cross at Point 3, the higher cost Machine A is more profitable and will continue to become more so as the sales volume rises.

It is evident that it's simply a matter of time until the increased profits on Machine A pay back the additional first cost of the more expensive machine. It should be noted that — depending on the relative productivity of new and old equipment — it is possible for a new machine to be more profitable than an old machine which has been fully depreciated and theoretically has no fixed cost (the dot-dash line in the graph).

Therefore, if a sufficiently high sales volume is reached (at Point 4 in this case), it is more profitable to dispose of a partly or fully depreciated machine and buy a new one, perhaps even the most expensive machine of several choices. ■

— Adapted from "Economics of Metal Stamping," by Robert F. Carlson, Torin Corporation. MM77-508.

PSQL An Economic Criterion For Minimizing Overall Inspection And Repair Cost*

by
Michael Z. Fruehwirth, Senior Development Engineer
Western Electric Co., Chicago, Illinois

INTRODUCTION

The primary functions of a quality control technique (system) are to assure that a certain quality level is maintained and to assure that this level is maintained at a minimum cost so that the producer's profit is optimal. Concern over the economic aspects of quality control dates back to the early stages of its development and to its founder, W. A. Shewhart[1], who states that, "the object of control is to enable us to do what we want to do within economic limits."

However, the economics of maintaining the desired quality level have not been sufficiently explored. A great number of procedures used in industrial quality control and assurance are based on sampling plans that maintain outgoing quality at a certain level. This level is determined in an arbitrary fashion based on the quality control practitioner's judgment, customer reaction, customer demand, process capability, potential loss of good will and other factors. Some economic models have been treated in literature,[3-9] but none of these can be generalized to fit all situations. This article proposes and develops a model that uses continuous sampling to minimize the cost of inspecting and repairing a multi-component unit produced in a continuous multi-stage process. The model incorporates sampling at a single intermediate point, and assumes a number of constraint conditions. Under these conditions, the model shows that an input quality level for which the expected cost of sampling at the intermediate point is equal to the expected cost of not sampling does exist at the intermediate point. This input quality level is defined as the Process Screening Quality Level (PSQL) and is shown to be independent of the sampling percentage. To make use of PSQL in a practical dynamic situation, a sampling procedure incorporating Dodge's single-level continuous sampling plan CSP-1 (See Reference 10) is then developed to minimize the overall inspection and repair cost.

DEVELOPMENT OF THE MODEL

The process for which the model is to be developed is assumed to be a continuous multi-stage operation producing a multi-component unit. In the process, it is assumed that one-hundred percent final inspection is specified to assume conformance to major requirements (electrical), that a defect in a single component will cause the defective unit to be routed to a repair station, and that repair is one-hundred percent effective. It is also assumed that equipment is available to perform inspection at a single intermediate point in the process.

The objective of the model is to establish an economic decision rule that will cause the removal and repair of defective units in such a way that the minimum cost will result.

PROCESS SCREENING QUALITY LEVEL

The criterion proposed in this article is based on mapping the decision region and determining the boundary between the two alternatives. For this determination, Figure 1, which shows that product flow in the two alternatives, will be used as a reference. The boundary in the proposed model will be called Process Screening Quality Level (PSQL)[8]. PSQL is defined as that input quality level at which the expected cost of process screening is equal to the expected cost of not screening at the single intermediate point.

When values of the input quality are above PSQL, in-process screening should be implemented to minimize the expected loss caused by defective units. The optimal policy maximizes the average fraction inspected $F'(\phi)$ above the PSQL consistent with the physical constraints, and meets the criterion of minimizing both the sampling and screening risks.

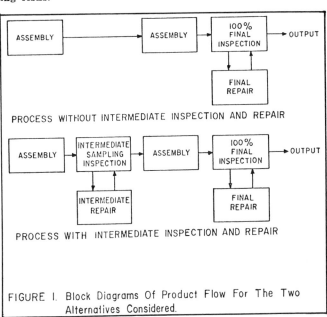

FIGURE 1. Block Diagrams Of Product Flow For The Two Alternatives Considered.

Screening and sampling risks are defined as follows: screening (detailing) risk is the probability that the product will be one-hundred percent inspected when the decision based on economic considerations should be to sample, and sampling risk is the probability that the product will be sampled when the decision based on economic considerations should be to one-hundred percent inspect.

Costs in the model are defined as:
c_1 = the cost of inspecting a unit at the intermediate point,
c_2 = the cost of repairing a defective unit at the intermediate point (this cost includes the cost of locating the cause of the defect, of repairing the unit, and of reinspecting the unit),
c_3 = the cost of inspecting a unit at the final point, and
c_4 = the cost of repairing a defective unit at the final point (this cost includes the cost of locating the cause of the defect, of repairing the unit, and of reinspecting the unit).

As additional components are added after the intermediate point, the unit becomes more complex. Thus, the costs of inspecting and of repairing a defective unit will normally be greater at the final point than at the intermediate point. Hence, $c_3 > c_1$ and $c_4 > c_2$.

Let d_1 represent the decision not to sample at the intermediate point and d_2 represent the decision to sample at the intermediate point. If d_1 were to be implemented (refer to top of Figure 1), the cost D_1 for inspection and repair of all units in a given lot of N units would contain the following cost components:

$C_{13} = c_3 N$ = the cost of inspection at the final point and
$C_{14} = c_4 N\phi$ = the cost of repairing defective units at the final point.

where: N = average number of units of product produced in a period
ϕ = incoming quality level.

[Continued on next page]

*Reprinted by permission of the American Society for Quality Control, Inc.

PSQL [Continued]

Thus,
$$D_1 = C_{13} + C_{14} = N(c_3 + c_4\phi) \quad (1)$$

If d_2 were to be implemented (refer to bottom of Figure 1) with the average fraction inspected equal to $F(\phi)$, the cost D_2 for inspection and repair of all units in a given lot of N units would contain the following cost components:

$C_{21} = c_1 N F(\phi) =$ the cost of inspecting at the intermediate point,

$C_{22} = c_2 N \phi F(\phi) =$ the cost of repairing defective units at the intermediate point.

$C_{23} = c_3 N =$ the cost of inspection at the final point, and

$C_{24} = c_4 N \phi [1 - F(\phi)] =$ the cost of repairing defective units at the final point.[9]

Thus,
$$D_2 = C_{21} + C_{22} + C_{23} + C_{24} =$$
$$N[c_1 F(\phi) + c_2 \phi F(\phi) + c_3 + c_4 \phi - c_4 \phi F(\phi)] \quad (2)$$

By equating the expected costs D_1 (equation 1) and D_2 (equation 2) of the two alternative decisions d_1 and d_2, the boundary of the cost indifference (ϕ is defined to equal ϕ^* at this boundary) can be mapped for the cost parameters developed above. After simplifying and rearranging the terms, the equations reduce to

$$0 = (c_1 + c_2\phi^* - c_4\phi^*) F(\phi) \quad (3)$$

Since $F(\phi) \neq 0$,

$$\phi^* = \frac{c_1}{c_4 - c_2} \quad (4)$$

Since $c_4 > c_2$ an equal cost for the two alternatives does exist and occurs when the input quality level at the intermediate point is numerically equal to the value given by equation (4) provided that $0 < \phi^* < 1$. If $\phi^* \geq 1$, screening at the intermediate point is economically infeasible. This level at the intermediate point is defined as the Process Screening Quality Level (PSQL). It can be seen from equation (4) that PSQL is independent of the value of $F(\phi)$.

By comparing equations (1) and (2) for levels above and below PSQL, decision d_1 is found to be lower in cost for $\phi < \phi^*$, and decision d_2 is found to be lower in cost for $\phi > \phi^*$.

Occurrence of the same and other types of defects at points subsequent to the intermediate stage is independent of the action taken at the intermediate stage. The majority of physical situations are of this type. In the few instances when this assumption does not hold the optimal ϕ^* (PSQL) would tend to increase.

THE CHOICE OF A SINGLE-LEVEL CONTINUOUS SAMPLING PLAN WITH PSQL

The basic difference between PSQL continuous sampling plans and previously developed continuous sampling plans is that the PSQL plans are used to control quality into the final inspection point to obtain minimum cost and, as such, do not have an Average Output Quality Limit (AOQL) as a measure of effectiveness. The AOQL does exist for such plans and can be determined in the usual manner.

CSP plans with PSQL are defined by the unique properties of the $F(\phi)$ curve for each plan. Ideally, the continuous sampling plan with PSQL would have a $F(\phi)$ curve as shown in Figure 2.

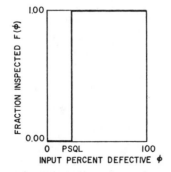

FIGURE 2. The Idealized Fraction Inspected Versus Input Percent Defective Curve.

However, such a sampling plan is an impossibility since this would mean "no inspection" when $\phi < \phi^*$ and "screening (100 percent inspection)" when $\phi > \phi^*$. Hence, a desirable plan should approximate this condition within given constraints. The choice will be influenced by such constraints as availability of trained inspection personnel and suitability of the process to the different rates of sampling. In practice these constraints generally provide a limit to the value of the sampling rate f for the CSP plan. This rate f and the value i (number of successive units required in clearing sequence) influence the dynamics of the CSP plan. These values in combination should maximize the sharpness of the $F(\phi)$ curve in the area of the PSQL to make the CSP plan with PSQL as close as possible to the ideal. The dynamics of the CSP plan with PSQL depend on the relative slope of the $F(\phi)$ in the area of PSQL. Within the limits of f and i, there exists a CSP plan that will maximize this relative slope and thus minimize the screening and sampling risks.

To provide guidelines for the selection of a CSP plan with PSQL, another definition is introduced. The indifference sample detail level (ISDL) of a single-level continuous sampling is defined as that input percent defective that requires examination of fifty percent of the product over the long run for a given sampling plan. It is designated as $F(\phi)_{0.5}$, and it is desired that this fraction inspected occur when the input percent defective is equal to the PSQL. If a limit on f (or i) is specified, the corresponding value for i (or f) for a CSP-1 plan with maximum slope passing approximately through $F(\phi)_{0.5}$ and the PSQL can be found from the following equation:

$$F(\phi)_{0.5} = \frac{f}{f + (1-\phi^*)^i (1-f)} = 0.5$$

where: $\phi^* = $ PSQL

Thus, an entire family of curves for various combination of f and i can be developed by specifying PSQL and ISDL. A single-level continuous sampling plan with PSQL can be completely defined by specifying f and i in addition to ISDL.

Since the detailing and sampling risks are equal to zero at the PSQL point, and since the cost of making a wrong decision in the vicinity of PSQL has been minimized by maximizing the slope in this area, the overall inspection and repair cost of a relevant process should be minimized.

A numerical example will now be considered to clarify this technique.

NUMERICAL EXAMPLE

Let the continuous multi-stage process that is constrained as indicated above be one that produces 1,000 units per week for 50 weeks a year. (N = 1,000 and M = 50). The costs of inspection and repair at the intermediate and final points are given by:

$c_1 = \$1.00$/unit = the cost of inspecting a unit at the intermediate point,

$c_2 = \$20.00$/unit = the cost of repairing a defective unit at the intermediate point,

$c_3 = \$1.75$/unit = the cost of inspecting a unit at the final point, and

$c_4 = \$30.00$/unit = cost of repairing a defective unit at the final point.

Based on these costs, equation (4) yields

$$\phi^* = \text{PSQL} = \frac{1.00}{30.00 - 20.00} = 0.1 \text{ or } 10 \text{ percent} \quad (6)$$

Based on the size of the production period, the availability of trained inspection personnel, the suitability of the process to the different rates of sampling, the average run, and the frequency of the defective product, the value of f might be chosen by the manufacturer to be 1/50. Equation (5) is then used to determine that i is approximately 37. The particular values for $F(\phi)$ can then be found from the following formula:

$$F(\phi) = \frac{1/50}{1/50 + (1-\phi)^{37} (1-1/50)} \quad (7)$$

The curve calculated from this formula is shown in Figure 3. The manufacturer has made studies of his process and has determined that the average weekly production run as a function of input percent defective is given in the long run by the curve $m(\phi)$ as shown in Figure 4.

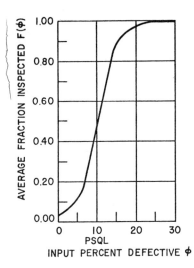

FIGURE 3. The Actual Average Fraction Inspected Versus Input Percent Defective Curve.

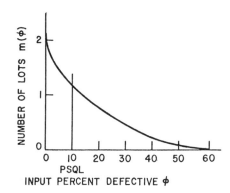

FIGURE 4. The Manufacturer's Number Of Lots Versus Input Percent Defective Approximation.

Three possible decisions for achieving the optimal economic solution for this example will be considered. The decisions at the intermediate point are:

1. Decision d_1 (no inspection) with cost
$$D_1 = N(c_3 + c_4\phi) \quad (8)$$
2. Decision d_2 (CSP plan with PSQL) with cost
$$D_2 = N[c_1 F(\phi) + c_2\phi F(\phi) + c_2 + c_4\phi - c_4\phi F(\phi)] \quad (9)$$
3. Decision d_3 (one-hundred percent inspection) with cost
$$D_3 = N(c_1 + c_2\phi + c_3)$$

The savings resulting from choosing decision d_2 instead of d_1 and from choosing d_3 instead of d_1 can be approximated by quantizing Figure 3 and 4 for integral values of ϕ and by using the following equation:
$$S = \sum_{\phi_1} m_1 D(\phi_1) \quad (11)$$
where:
S = the average total savings in M lots,
m_1 = the number of times in M lots that the input percent defective is equal to ϕ_1, and
$D(\phi_1) = D_1(\phi_1) - D_2(\phi_1)$ = the amount saved (lost if negative) by intermediate sampling for a single lot that has an input percentage defective of ϕ_1.

The payoff table for d_2 instead of d_1 is shown as Table I. A similar table could be obtained for d_3 instead of d_1, and would give a resulting average savings of $37,732.60. Thus, comparison of the tables shows that the CSP-1 plan with PSQL does minimize the cost over any other alternative provided by the assumed model.

TABLE 1—COMPARISON OF COSTS D_1 & D_2

ϕ (Percent)	m	$F(\phi)$	D_1-D_2 (Dollars)	mD_1-D_2 (Dollars)	ϕ (Percent)	m	$F(\phi)$	D_1-D_2 (Dollars)	D_1-D_2 (Dollars)
00	2.016	.0200	−20.00	−40.32	31	.705	1.0000	2100.00	1,480.50
01	1.955	.0280	−25.20	−49.27	32	.685		2200.00	1,507.70
02	1.915	.0414	−33.12	−63.42	33	.645		2300.00	1,483.50
03	1.854	.0593	−41.51	−76.96	34	.623		2400.00	1,500.00
04	1.814	.0845	−50.70	−91.97	35	.584		2500.00	1,450.00
05	1.754	.1198	−59.90	−105.06					
06	1.713	.1667	−66.68	−114.22	36	.564		2600.00	1,466.40
07	1.653	.2301	−69.03	−114.10	37	.524		2700.00	1,414.80
08	1.572	.3096	−61.92	−97.32	38	.504		2800.00	1,411.20
09	1.532	.4008	−40.08	−61.40	39	.483		2900.00	1,400.70
10	1.491	.5013	0	0	40	.443		3000.00	1,329.00
11	1.451	.6018	+60.18	+87.32	41	.423		3100.00	1,311.30
12	1.411	.6987	139.74	+197.17	42	.383		3200.00	1,255.60
13	1.370	.7780	233.40	+319.76	43	.362		3300.00	1,194.60
14	1.330	.8430	337.20	448.48	44	.322		3400.00	1,094.80
15	1.290	.8909	445.45	574.63	45	.302		3500.00	1,057.00
16	1.250	.9173	556.38	695.47	46	.282		3600.00	1,015.20
17	1.209	.9533	667.31	806.78	47	.221		3700.00	969.40
18	1.169	.9645	771.60	902.00	48	.221		3800.00	839.80
19	1.129	.9841	885.69	999.94	49	.201		3900.00	783.90
20	1.088	.9874	987.40	1074.29	50	.181		4000.00	724.00
21	1.048	.9920	1091.20	1143.58	51	.161		4100.00	660.10
22	1.008	.9951	1194.12	1203.67	52	.141		4200.00	592.20
23	.967	.9968	1295.84	1253.07	53	.120		4200.00	516.00
24	.940	.9999	1399.86	1315.87	54	.100		4400.00	440.00
25	.908	1.0000	1500.00	1362.00	55	.080		4500.00	360.00
26	.866		1600.00	1385.60	56	.060		4600.00	276.00
27	.846		1700.00	1438.20	57	.040		4700.00	188.00
28	.806		1800.00	1450.80	58	.020		4800.00	96.00
29	.766		1900.00	1455.40	59	.004		4900.00	19.60
30	.745	1.0000	2000.00	1490.00	60	.000	1.0000	5000.00	0
							Total Savings		46,620.87

CONCLUSION

This article has proposed a new application for continuous sampling plans in minimizing the cost of in-process inspection and repair. By establishing the existence of a level of input quality PSQL for which the cost of inspecting at a single intermediate point is equal to the cost of not inspecting and by relating this PSQL to Dodge's CSP-1 plan through the use of the Indifference Sample Detail Level, an economic model has been generated for a multi-stage manufacturing process. Within the constraints assumed, the model controls the quality at a final inspection point so that minimum overall inspection and repair cost occurs.

REFERENCES

1. Fruehwirth, M. Z., "An Economic Criterion for Minimizing Overall Inspections and Repair Cost," **Western Electric Engineer**, Vol. XIV, No. 2, April, 1970.
2. Fruehwirth, M. Z., "PSQL An Economic Criterion for Single Level Continuous Sampling Plans," Unpublished M.S. Thesis, Illinois Institute of Technology, Chicago, Illinois, 1968.
3. Guthrie, D., Jr. and M. V. Johns, Jr., "Bayes Acceptance Sampling Procedures for Large Lots," **Annals of Mathematical Statistics**, Vol. 30, 1959.
4. Hald, A., "The Compound Hypergeometric Distribution and A System of Single Sampling Inspection Plans Based on Prior Distribution and Costs," **Technometrics**, Vol. 2, No. 3, August, 1960.
5. Hamaker, H. C., "Some Basic Principles of Sampling Inspection by Attributes," **Applied Statistics**, Vol. 7, 1968, pp. 149-159.
6. Heeremans, J. H., "Determination of Optimal In-Process Inspection Plans," **Industrial Quality Control**, Vol. 8, 1962, pp. 22-37.
7. Shewhart, W. A., "Economic Control of Quality of Manufactured Products," D. Van Nostrand Company, Inc., 1931.
8. Smith, B. E., "Some Economic Aspects of Quality Control," Technical Report No. 53, Applied Mathematics and Statistics Laboratories, Stanford University, 1961.
9. Smith, B. E., "The Economics of Sampling Inspection," **Industrial Quality Control**, Vol. 22 (March 1965), pp. 453-458.
10. Dodge, H. F., "A Sampling Inspection for Continuous Production," **Annals of Mathematical Statistics**, Vol. 14, 1943, p. 264-279.
11. For more detailed references see bibliography of the reference no. 2.

ESTIMATING OF MANUFACTURING JOINT COSTS

By Adam Malolepszy
Senior Manufacturing Engineer
AMF—Head Division
Boulder, Colorado

Phillip F. Ostwald
Professor of Mechanical and Industrial Engineering
Department of Mechanical Engineering
University of Colorado
Boulder, Colorado

This paper initially reviews the definition of joint cost from the perspective of the manufacturing cost estimator. Continuing, cost traceability leading to the split point is discussed. Traditional and cost estimating approaches are distinguished. Finally, methods for handling joint product cost are identified for an elementary problem.

INTRODUCTION

A common function of a manufacturing engineer is to prepare a detailed cost analysis of a manufacturing process or procedure. The estimate identifies the true cost of converting a raw material into a finished product. This information may determine the profitability of a proposed product and stop or promote it. Traditional approaches to joint costs of joint products involve concepts which serve accounting or marketing needs and do not necessarily reflect the true or actual costs of specific interest to the engineer. However, it is possible to identify joint costs of joint products and determine the exact costs by using fundamental concepts.

JOINT PRODUCTS AND JOINT COSTS

Joint products are those products which result from the processing of a singular raw material supply. Joint products remain joint up to the point at which the products are divided into separable entities. The point of division is called the "split point". Material and labor costs up to the split point are referred to as joint costs.

The key element in this definition is the concept of the singular raw material which by virtue of a processing step becomes two or more discrete products. For example, the processing of raw milk into cream and skim milk illustrates the conversion of singular raw material into two discrete products which will be individually refined and marketed.

Another example is the rough log which upon sawing and milling becomes 1st grade lumber, 2nd grade lumber and sawdust. A molded plastic part where the die has several cavities is a common production joint cost problem. Similar situations occur in a variety of industries, but especially in process oriented industries such as chemical, petroleum, food, metallurgical or timber. These industries include processes to make marketable products of what were essentially by-products of basic processing steps. The problem in processes which result in multiple products is the tracing of the cost contribution of raw material and labor to the individual final products. Often the choice of allocators is not clear, or product lines are not direct, or other factors appear to prevent cost traceability.

TRADITIONAL APPROACHES, ACCOUNTING AND MARKETING

Accounting concerns with joint product costs essentially revolve around the task of distributing raw material and labor costs incurred prior to the split point in an equitable and easy manner to the products produced. Marketing strategy can affect allocation schemes by requiring that some products subsidize others. This practice recognizes that selling price or market value of a product is not necessarily proportional to processing costs. The combined influence of accounting and marketing philosophy on allocation of joint costs to joint products may hide or distort real or true costs.

The allocation of joint cost on the basis of quantity can generate cost data which may be invalid for accounting purposes. This arises when products are valued for inventory purposes at levels higher than their possible selling price. A similar situation occurs when marketing policy dictates price levels which do not reflect manufacturing costs, but rather selling costs, or promotional costs. Often the profitability of one product cannot be gauged by available data, whereas, the profitability of a group of products may be properly represented due to schemes of cost accounting. The reverse may also be true.

In the previous example of a rough log subject to a splitting process, the production of first grade lumber and other lumber products are not independent actions, as the production of first grade lumber is necessarily associated with the production of lesser grade lumber and sawdust. The processing investment to finish and market secondary lumber products is often subsidized by profits (or operating costs) from first grade lumber production. Profitability decisions on specific products cannot be made on the basis of accounting data, which may cover a group of products, and not reflect the true costs of an individual product. Herein arises the need for an exact cost estimate, an engineering cost estimate which accurately examines joint product costs. We point out, however, that the traditional approaches are justified for their intended purpose.

DISTINCTIONS BETWEEN DISTRIBUTING AND CONVERTING JOINT COSTS

A distinction is necessary between the distributing and converting types of joint costs. The distributing type of joint costs is illustrated by the multiple-cavity die problem as the plastics pellets are intermingled prior to the molding operation. The characteristic of the plastic is not altered (in a joint cost sense) and cost traceability is valid albeit complicated. An enlarged example describes the techniques of the distributing type of joint cost shortly. In processing industries a specific quantity of raw material (rawterial) is transformed into a singular new product or material. A simple example is water + heat process = steam. Splitting of rawterial requires that the essential cost nature of the rawterial be changed, resulting in two or more discrete products or materials with differing characteristics and physical measures or values. Figure 1 describes this distinction.

MANUFACTURING CASE PROBLEM IN A DISTRIBUTING TYPE OF JOINT COST

Consider an injection molding material which is blended, pigmented and molded. Assume that a molded part is one piece in a multiple cavity mold, and the die has runners that connect the pieces from a common sprue. Furthermore the operator tends two machines which are operating at different cycles, and each machine produces a different part, designated A and B. Set A and B are three and four cavity molds respectively. The essential design features are given by Figure 2.

At this point it is convenient to consider allocators, or those units of reference which are commonly used to prorate or allocate costs. In an estimate of costs, the essential function of the word "cost" is the establishment of a unit of valuation per unit of reference. Commonly this is a dollar value per unit of measure ($/ft^2, $/lb, $/KW, $/hr, $/product unit, etc). In the event that a cost allocation is made, (to meet accounting, engineering needs) it is useful to think of fundamental reference units which are readily measured. The simpler the measured unit, the clearer its use becomes, and the more accurate are the conclusions drawn from the analysis. Allocators can be classified as follows:

a) Physical measure (geometry, weight, shape, etc.)

b) Energy (BTU, KW, etc.)

c) Time (second, year, man-hour, etc.)

d) Units (of finished product, each, 100 units, etc.)

In general, multiple allocators are redundant (even though to properly define the usage of energy, a term such as KW-Hr. can be used). For traceability of costs the unit of reference needs to be defined to avoid changing the unit through a process. For example, if lb. is used in the rawterial stage, lb. is convenient as a finished product, rather than ft^2. This becomes important with split converting where split products may be in different physical states. Unnecessary unit conversion should be avoided. For example, rather than defining the products of a petroleum cracking process as gallon or barrel and ft^3, a common unit such as lb. can be used.

A problem confronting a cost estimator is given as: "Two machines are molding round preforms or buttons of a flour-like plastic. The rawterial costs $2.50/lb. and has a density of 0.0275 lb./in.^{3}. Machine A and B have a production rate of 400 and 300 sets per hour. A set is composed of 3 or 4 buttons having a sprue and runners. Dimensions of the set from Machine A and B are given in Figure 2. The runners and the sprue are considered waste and are eventually trimmed as only the button is used. One operator, who is paid $5.50/hr., tends machine A and B." Can you, the reader, solve this problem? Four approaches have been isolated in Table 1, and there may be others, but these four are more reasonable than others. Assume that the waste is neutral to the cost of the problem, i.e., it has neither a credit or deficit to the value of the final product.

In this case problem, the material cost is uniformly related to measured values of output. Labor input, however, can be apportioned on four basis. These basis arithmetically lead to differing solutions for the contribution of labor cost to joint cost. Four solutions for designs A and B have been isolated on Table 2. Which one of the four is best? The answer is a knotty one at best, and perhaps, we sidestep the issue by affirming that details outside of the calculations will choose the "best" one. Cost traceability is the important element in this example. Despite the apparent complications of this problem, the costs of the finished product can be traced to the original rawterial using a simple allocator basis and measure of quantity. This element of direct cost traceability is characteristic of simple processing of rawterials.

However, in a splitting process, the converting of a rawterial creates two or more new products, in a constant or variable ratio, and with relative values and quantities disproportionate to original rawterial values and quantities. The element of direct cost traceability becomes lost if the essential quantity measures vary or become meaningless. This is characteristic of true joint costs of joint products of converting processes. By following certain rules these difficulties can be circumvented and an engineering cost estimate can be calculated.

CONVERTING-TYPE JOINT COSTS

When the converting type of joint costs occur, the key to unscrambling the cost allocation is the selection of a primary product in the company or in the processing group. With a selection made, unit costs for processed rawterial (proterial) can be established. The primary product is the product which forms the financial and physical justification for a company or division or process to exist. All other products are secondary products, regardless of their value, and would not exist were it not for the production of the primary product. Where the primary product cannot be immediately identified, or can be changed by minor process changes, an economic profitability analysis should be performed. For example, in the dairy industry, raw milk can be processed into cream, skim milk, powdered milk, cheese, and butter. All these products can be produced, yet the operation may be set up to optimize the production of only one product, say cheese. If concentrating on cheese production optimizes the profitability of the product line, then cheese is the primary product. The identification of a primary product allows the process to be presented schematically as a direct flow from rawterial to finished product with all secondary products branching off at their split points. The flow of material is an engineering decision, and once a design has been chosen cost analysis can begin. Converting industries unlike the manufacturing, fabrication, and the durable goods industries, deal with this kind of a joint cost problem.

CONCLUSIONS

Joint costs are frequent problems faced by the cost estimator. In manufacturing, rawterial is split and a unit of reference chosen to have an accurate, sensible, and easily traced joint cost.

TABLE 1. STRUCTURE TO JOINT COST PROBLEM

Facts for the designs:

A. Volume = 3.6816 in.^{3} for 3 buttons
    Weight = 0.1012 lb. for 3 buttons
    Sprue   = 1.5708 in.^{3}, weight = 0.0432 lb.
    Runners = 0.1473 in.^{3}, weight = 0.0041 lb.

B. Volume = 1.5708 in.^{3} for 4 buttons
    Weight = 0.0432 lb. for 4 buttons
    Sprue   = 1.5708 in.^{3}, weight = 0.0432 lb.
    Runners = 0.1964 in.^{3}, weight = 0.0054 lb.

Set A	in.^{3}	lb.	$
Total shot	5.3996	0.1485	0.3712
Waste	1.7181	0.0432	0.1080
Good product	3.6816	0.1012	0.2531

Unit material cost for Set A = $\frac{0.3712}{3}$ = $0.1237/unit

Material in good product for Set A = $\frac{0.2531}{3}$ = $0.0844/unit

Set B	in.^{3}	lb.	$
Total shot	3.3380	0.0918	0.2295
Waste	1.7672	0.0486	0.1215
Good product	1.5708	0.0432	0.1080

Unit material cost for Set B = $\frac{0.2295}{4}$ = $0.0574/unit

Material in good product for Set B = $\frac{0.1080}{4}$ = $0.0270/unit

Set A output = 400 (0.1485) = 59.3960 lb./hr
Set B output = 300 (0.0918) = 27.5400 lb./hr.
Total for Machine A and B = 86.936 lb./hr.

1. Labor $5.50/hr. apportioned on total lb./hr. basis

   Labor rate for Machine A = $\frac{59.396}{86.936}$ (5.50) = $3.758/hr.

   Labor rate for Machine B = $\frac{27.540}{86.936}$ (5.50) = $1.742/hr.

2. Labor $5.50/hr. apportioned on shots/hr. basis and total shots = 700.

   Labor rate for Machine A = $\frac{400}{700}$ (5.50) = $3.143/hr.

   Labor rate for Machine B = $\frac{300}{700}$ (5.50) = $2.357/hr.

Table 1 (continued)

3. Labor $5.50/hr. apportioned on units/hr. basis

$$\text{Machine A units} = 1200 \text{ units/hr.}$$
$$\text{Machine B units} = 1200 \text{ units/hr.}$$

$$\text{Labor rate for Machine A} = \frac{1200}{2400} \times 5.50 = \$2.75/\text{hr.}$$

$$\text{Labor rate for Machine B} = \frac{1200}{2400} \times 5.50 = 2.75/\text{hr.}$$

4. Labor $5.50/hr. apportioned on lb./hr. of final product produced

$$\text{Total lb./hr. for Machine A} = (\frac{1200}{3})(0.1012) = 40.48 \text{ lb./hr.}$$

$$\text{Total lb./hr. for Machine B} = (\frac{1200}{4})(0.0432) = 12.96 \text{ lb./hr.}$$

$$\text{Labor rate for Machine A} = \frac{40.48}{53.44} (5.50) = \$4.166/\text{hr.}$$

$$\text{Labor rate for Machine B} = \frac{12.96}{53.44} (5.50) = \$1.334/\text{hr.}$$

TABLE 2. SOLUTIONS TO JOINT COST PROBLEM

Design	Basis for Apportioning of Labor Cost	Material	Labor	Unit Joint Cost of Labor + Material
A	lb./hr. input	$0.1237	0.0031	$0.1268
A	shots/hr.	$0.1237	0.0026	$0.1263
A	units/hr.	$0.1237	0.0023	$0.1260
A	lb./hr. output	$0.1237	0.0035	$0.1272
B	lb./hr. input	$0.0574	0.0015	$0.0589
B	shots/hr.	$0.0574	0.0020	$0.0594
B	units/hr.	$0.0574	0.0023	$0.0597
B	lb./hr. output	$0.0574	0.0011	$0.0585

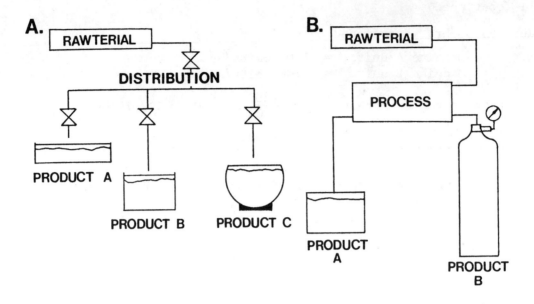

Fig. 1. Two classes of joint cost. (A) Rawterial being "distributed" and (B) Rawterial being "converted" into two or more products.

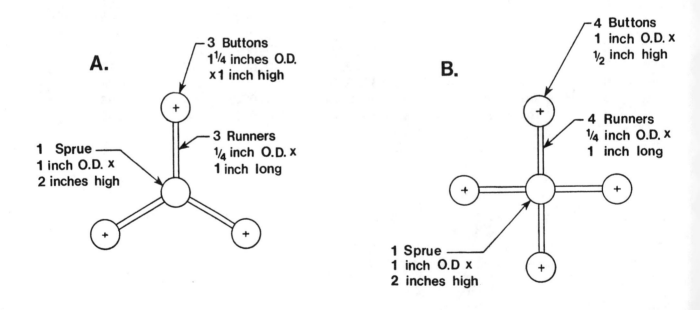

Fig. 2. Two designs for a distributed joint cost problem.

Computerized Standard Data—
A Powerful New Tool For Manufacturing Engineering

By **Leroy H. Lindgren**
Principal
and
Romeyn D. Murphy
Associate Consultant
Rath and Strong, Incorporated

Computerized Standard Data (CSD) is a computer software package that provides for the computation, documentation, and maintenance of production operation method and time standards. The implementation of CSD results in a powerful computer system with the potential for substantially improving the method and time standards analysis function.

INTRODUCTION

The major steps involved in the utilization of CSD are briefly summarized as follows:

1. The methods analyst performs his analysis of the production operation while recording the required descriptive, dimensional, and frequency related data on the input documents provided.

2. This information is keypunched and input to the CSD Methods Generation Module, resulting in a computer-generated operation method and time standard. This standard is provided in both printed form, for audit and shop production purposes, as well as in magnetic disk or tape storage form.

3. Any required changes or corrections can be easily made through single line or field additions, deletions, and changes, utilizing the CSD Methods File Maintenance Module along with the magnetic storage form of the original methods analysis. The resultant changes bring about the complete regeneration and printout of the updated operation method and time standard.

In the remaining sections of this article, background information, basic definitions, and examples are presented to provide a conceptual understanding of CSD, followed by an analysis of advantages from two separate viewpoints-- engineering and systems.

PROBLEM BACKGROUND

A top priority engineering function within many companies consists of the establishment of the specific method of manufacture for a given product, along with the labor and/or machine time standards for performing each operation of this method. Most engineering and management people will readily agree on the importance of the engineered method for the performance of a given production operation. However, unless a company is involved with labor incentives or is blessed with an unusually advanced thinking management, the need for the accurate operation time standard is often overlooked. A few of the more significant contributions that are provided by good labor time standards are the following:

1. <u>Engineering decisions based on true economics</u>

    A major factor in the evaluation and comparison of alternate facilities and methods of manufacture is an accurate estimate of the amount of time required to do the job for each of the alternatives.

2. <u>More accurate product pricing decisions</u>

    Without accurate labor standards one must depend on labor "guesstimates", multipliers, factors, and other devices that attempt to arrive at a new product price based on past average labor costs. Due to compensating errors this average labor cost may be fine for a product group taken as a whole, but way out of line for many of the individual items within the group.

3. <u>More successful production planning</u>

    One of the key ingredients for successful production planning is an accurate measure of the amount of time that will be required for each operation of those product items currently within the production load mix.

4. <u>Better capacity planning projections</u>

    Planning for future production capacity and facilities requirements is highly dependent upon both a good forecast of what future business will consist of, as well as an accurate measure of the impact of that forecast on existing and proposed facilities. This impact must be expressed in terms of demands on total available production time.

5. <u>Successful management performance evaluation</u>

    Evaluation of the performance of a production group within a company requires a common denominator for valid comparisons with other groups. Accurate labor time standards are a major contributor to the validity of the evaluation, without which a performance measuring system has little chance of success.

Perhaps the main reason that many companies do not have accurate time standards is with the utilization of conventional methods they are very expensive to arrive at and difficult to maintain. The determination of these time standards normally utilizes one or some combination of the following techniques:

1. Time Study

2. Predetermined Time Standards

3. Standard Data

The standard data technique involves the application of precisely constructed predetermined time values that are associated with specific blocks of work, referred to as macro work elements or Master Standard Data. These macro work elements are structured within logically segmented application sets in such a manner that various combinations of elements may be selected and combined to build the time standard for any member of a related group of work operations.

The main advantage of standard data, as compared to either the

time study or predetermined time standards technique, is the
elimination of much of the repetitive detailed analysis, resulting in greater consistency and accuracy of application
with a smaller investment in skills and effort than would
otherwise be required.

However, even with the utilization of the standard data
technique, a considerable amount of analysis and computational
effort still remains within the operation analysis function.
This requires a group of well trained, experienced engineering
analysts--a group which is difficult (if not impossible) to
keep fully staffed, due to the inherently high turnover rate
among the more qualified people. Further compounding the
problem is the constantly increasing operation analysis work
load being experienced by many companies, often due to a trend
toward shorter product life causing the more frequent occurrence of new product introduction and existing product
change. As a direct result, the methods and time standards
analysis function is forever becoming a more costly, time
consuming, and overburdened effort.

Recognition of these problems by Rath & Strong instigated
preliminary probings into various solution areas, eventually
resulting in the selection of the computer as a potentially
significant method and time standards application tool.
During initial investigation of the feasibility of a computer
approach, it became quite evident that the computer has been
historically under-utilized by manufacturing engineering.
This apparently is not usually due to any lack of desire on the
part of engineering, but often the result of a low priority
assignment from that group within most companies whose function
it is to plan the future utilization of its computer facility,
the Systems Group. This low priority often results from a
failure to consider the requirements of other areas of a
company, and how they blend in with the needs of manufacturing
engineering.

Recognition of this past failure resulted in an early development stage broadening of objectives to include a "total systems"
approach. Because of this, the resultant computer system
package is not only of great potential benefit to the Engineering Group, but also to the Systems Group for utilization
within both the existing framework and future design of their
Management Information System.

WHAT IS CSD?

To the manufacturing engineer, Computerized Standard Data is a
computer system to both aid with the analysis, computation, and
documentation of a production operation method and time standard,
as well as provide the capability for maintaining that analysis
with a minimum of effort. However, to the Systems Group, CSD
is a data entry and file maintenance system for creating and
maintaining a manufacturing information data base. This data
base, referred to as the Methods File, consists of all the
analysis input and descriptive information provided by the
manufacturing engineers for computer generation of the method
and time standards.

The Computerized Standard Data System consists of a total of
four computer programs. Two of these programs are responsible
for the initial edit and load of the above-mentioned Methods
File and periodic loading of the Element File (refer to
Figure 1). This Element File consists of all the standard data
macro elements with their respective standard time values. The
two remaining programs perform the methods generation and file
maintenance functions as described in the following sections
(refer to Figure 2).

Methods and Standards Generation

The Methods and Standards Generation Program sequentially
searches the Methods File to determine which specific pro-

duction operations have been newly added or updated by the Maintenance Program. Each operation that is found to be in this category is immediately generated, performing all required computations and standard data macro element retrievals. The results are printed out in the form of a document capable of providing both the engineer with detailed backup information, as well as the manufacturing organization with the required production information.

An example of the input analysis and resultant computer output reports for a typical manufacturing machining operation are displayed in Figures 3 through 5. This is one of several operations to be performed on this particular part. The manufacturing engineering analysis began with the entry of information on the input form in Figure 3. The first four lines relate to heading information, while the remaining lines are for the entry of detailed machine cutting information used in the Methods and Standards Generation Program for computation of the machine cutting time.

Figure 4 consists of a checklist of the standard data macro elements for defining the manual handling requirements across a particular group of turret lathes. The manufacturing engineering analysis has resulted in the selection of specific elements applicable to the operation under study. These were circled on the checklist and their frequency of occurrence was determined and entered. There is also a similar checklist of standard data macro elements for establishing the machine setup element occurrences, which is utilized in the same manner as the handling analysis form.

After the engineer has completed his analysis and entered all required information, the forms are submitted to the Data Processing Organization for key punching, verification, and computer processing. Figure 5 consists of the resultant computer output report intended for engineering backup information requirements. There is also a suppressed version of this report for the Manufacturing Organization, consisting of the left-hand half of the engineering report. A two-part segmented carbon output form is utilized, resulting in a computer processing cost-saving from printing both documents simultaneously on one computer line printer.

The input form in Figure 3 has also been designed to provide for the analysis and entry of information which is not directly related to a machine. An example of this would be an entirely manual assembly operation requiring the entry of descriptive instructions documenting the specific assembly method. This would utilize the entire input document line for pure descriptive information, ignoring all field headings, through selection and entry of appropriate code characters in column 79 of the Machine Cutting Data section. This information will then print on the output report exactly as entered on the input document line; and depending on the code selected, it will leave a specific number of spaces between lines. The various code characters available provide a high degree of flexibility, giving the engineer the capability for controlling the appearance of each individual output report.

The input form in Figure 4 could also be utilized for the analysis of assembly-type operations, providing a checklist of the standard data macro elements for determining the assembly time for a group of similar products. A similar checklist could also include job and workplace preparation elements for an assembly setup.

Methods File Maintenance

The Methods File Maintenance Program performs the logic required to add to the Methods File an entirely new analysis for a particular production operation on a specified part. This program also provides the engineer with a powerful tool for

alteration of an existing method through the insertion of
corrections and additions to an operation analysis already
existing on the Methods File. The input to this Maintenance
Program is in the form of Maintenance Transaction Cards consisting of the appropriate information along with a Maintenance
Code Letter indicating which of the following options has been
selected:

1. An entirely new analysis line may be added in any
   desired sequence position of the analysis for a
   production operation that exists on the Methods File.

2. An entirely new analysis line may be used to replace
   any existing analysis line.

3. Any existing analysis line may be deleted from the
   Methods File.

4. One or more data items within any existing analysis
   line may be changed while not affecting the remaining
   data items for that line.

5. An entire methods analysis for an operation of a
   specific part number may be deleted from the Methods
   File with one single Maintenance Transaction Card.

6. A single card entry is available to request the
   generation and printing of the output report for any
   existing operation of a part number on the Methods
   File.

The input forms for creating these Maintenance Transaction
Cards, displayed in Figure 6, are very similar in function to
their input form counterparts in Figures 3 and 4. The procedure for utilizing these forms involves the entry of the part
number, operation number, machine group, and line number for
that particular analysis line that is to be added or updated.
Following this, the actual information is recorded in the
proper fields along with the desired Maintenance Transaction
Code Letter.

After all of the necessary maintenance transactions have been
entered, the forms are submitted to data processing for key
punching, verification, and computer processing. The resultant output reports consist of documents for only those
operations that have been either changed, newly added, or
requested for printout.

MANUFACTURING ENGINEERING ADVANTAGES

Some of the obvious advantages provided by CSD for the manufacturing engineering function are as follows:

1. An approximate 20 per cent reduction in the total
   methods analysis and maintenance manpower investment.

2. Increased accuracy and consistency of application.

3. More currently maintained methods and time standards
   through simplification of the change introduction
   process.

4. Elimination of the typing bottleneck normally
   associated with issuance of new and changed methods.

In addition to these advantages from the basic CSD system,
there are many potential advantages to engineering that are a
direct result of the establishment of a CSD data base. The
importance of having this highly detailed information in the
computer-processable storage medium provided by the Methods
File eventually should far outweigh the advantages previously
discussed. Most of the items mentioned in the following

sections would only require fairly simple computer programs as modular additions to the basic CSD system.

1. **Computer Macro Element Change**

    Many companies experience frequent changes that directly affect the basic time value for a standard data macro element. This type of change would normally require the extremely costly changing of every operation time standard that uses this particular element value, very often numbering in the thousands. With CSD, this would involve the manual alteration of only one time value on the Element File, followed by a completely computer handled search and regeneration of every operation method affected.

2. **Computer Audit Trail Creation**

    Usually some form of backup information is required to provide an audit trail covering past changes to operation method and time standards, often involving the cumbersome and space-consuming filing of all previous revisions. A modular program addition to CSD could provide an audit trail through computer storage of method analysis lines that are affected by each revision. This would allow for the re-creation of an operation at any specified revision level, through a reversal of the previously described maintenance process.

3. **Computer Procedural Control**

    Various control procedures can be easily programmed in to CSD to provide absolute control of such areas as temporary method and time standards, temporary allowances, and release of changed operation standards. This would provide more effective control of these and other critical areas at a much-reduced expense than is normally involved with manual controls.

4. **Computer Alternate Machine and New Part Analysis Generation**

    Where the analysis of an operation by the manufacturing engineer involves the selection of a specific machine from a group of similar machines, he attempts to select the best machine for the given part. However, when the part is to be actually manufactured, it may have to be scheduled for an alternate machine requiring a similar analysis to develop a method and time standard for the operation in question. A fairly simple program modular addition to CSD could provide the engineer with a duplicate of the original machine analysis, in the form of an addition to the Methods File. Utilizing the Methods File Maintenance Program, the manufacturing engineer would then only have to make whatever changes that are required to tailor the analysis specifically to the alternate. Program modifications could also be made to compute the labor and machine cost variance for utilization of this alternate machine in place of the optimum.

    A similar approach could be utilized for prefabrication of an analysis for an entirely new job, based on a previously constructed analysis for a similar job. This would involve computer insertion of the new part number, operation number, and machine number, combined with computer duplication of the original analysis onto the Methods File. The engineer could then utilize the Methods File Maintenance Program for specifically adopting the duplicate analysis to the new job.

All of these benefits are directly attainable from the currently available CSD system as described in the previous section. A more advanced version of CSD will provide added benefits to the manufacturing engineer. This version will have two new input files of information as part of the CSD data base. The first is a Facility File, providing the actual speeds, feeds, horsepower ratings, and other information for each manufacturing facility group within a manufacturing plant. The second file is a Tool File to provide functional, physical, and dimensional information for all standard and special tooling utilized within the manufacturing facilities.

In addition to these two files, new tables and logic have been included within the Methods and Standards Generation Program. As a result, computer calculation and selection of actual facility speed and feed settings will be provided. The net effect of this further sophistication of CSD is that less input on the part of the manufacturing engineer will result in a more extensive analysis, specification, and documentation of the production operation method and time standard.

SYSTEMS ADVANTAGES

CSD supplies the Systems Group with a vast amount of accurate and timely manufacturing information, highly unique in the level of detail provided. Because of its extreme detail, this data base has the potential for providing substantial cost savings through the application of Group Technology theory. This involves the analysis and comparison of the total product manufacturing details, the grouping of previously unrecognized similarities, and the ultimate removal of much of the uniqueness and seemingly miscellaneous nature within each group. Current product mix may then be grouped for production purposes based on these similarities in setup, configuration, skill level, and work requirements, with the objective of decreasing setup and learning costs, while increasing throughput and optimum utilization of facilities.

Whether the full potential of this detailed information is exploited or not within a given company, CSD can provide the Systems Group with an extensive amount of basic information for more common uses, among which are the following:

1. Shop Methods File for computer generating detailed shop work orders for production lots.

2. Job Routing File for providing the sequence and machine center or department number of each operation for a given production order.

3. Machine, labor, and tooling requirements for planning purposes utilizing a production forecast.

4. Data extraction of information for other systems, among which could be included:

    - standard cost data

    - tooling information

    - labor grade or skill level requirements

    - planning quantities

    - materials

    - miscellaneous descriptive information

This data provided by CSD has one distinct advantage: it fulfills, to a rarely attained degree, the extremely important requirements commonly shared by all Management Information Systems--accurate and timely input information. Without this, the value of an MIS system is severely limited and, in fact, its actual survival may be endangered. Among the major causes for inaccurate and untimely input information are the following:

1. The task of physically providing the actual data has been subdelegated to a very low level, resulting in a lack of sufficient knowledge on the part of the input provider.

2. The input provider does not receive any relevant feedback of useful information for accuracy checking.

3. The input data provided gives no direct benefits to the supplying organization. Depending upon the current political climate, this organization may be offering a high degree of "lip service" without actually having any of the necessary motivation and interest.

4. The entire data collection system has been superimposed on an existing manual system in an unnatural manner, requiring extraneous and burdensome efforts on the part of the information supplying organization.

As a result of these data collection system design failures, rigid controls have to be provided along with constant vigilance by the Systems Group, often leading to even greater distaste for the entire system and further aggravation of the problem. In light of this, the advantages of a CSD-maintained data base are fairly evident:

1. The CSD data base is built and maintained by manufacturing engineers. There is probably no group with more production knowledge within a typical company.

2. The engineers are provided with a feedback document of highly relevant information, the accuracy of which is of great importance to the individual who supplied the information.

3. The Engineering Organization has a vested interest in CSD, providing true motivation for insuring accurate input and proper use of the system.

4. The CSD System is an ideal replacement for the manually performed function, serving as a tool for relieving the burdensome clerical and computational efforts now required.

CONCLUSION

For better than two decades engineering specification has been reconciled to modest improvements in existing technology. Many of these improvements have been based on high-level objectives; however, the critical procedures for detailed methods and maintenance proved to be extremely cumbersome. CSD and Group Technology may provide the breakthrough, not only in performing the method analysis and maintenance function at lower cost, but also in achieving an improvement in manufacturing costs.

Engineering management is directly responsible for taking the initiative of exploring these potential benefits for manufacturing engineering. This investigation should be conducted utilizing a "systems approach" for examining both the direct advantages available to engineering, as well as the potential advantages to the entire company from better information and management control. The results of this study could then be utilized for a complete analysis and review of the future role reserved for the manufacturing engineering contribution to the company Management Information System.

If engineering management does not take this initiative, eventually someone else will; and they will most likely be from outside the Engineering Organization. This has historically led to a very different involvement with the company information system, demanding much from the engineer, while providing very little in return.

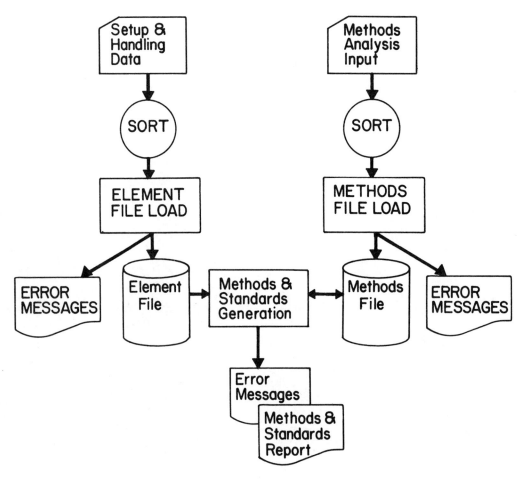

Figure I
THE INITIAL LOADING OF THE METHODS AND ELEMENTS FILES

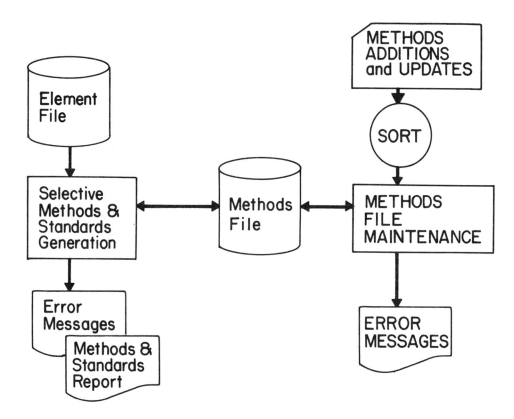

Figure 2
NORMAL MAINTENANCE OF THE METHODS FILE, INCLUDING GENERATION OF NEW OR CHANGED METHODS AND STANDARDS REPORTS.

MACHINING AND TIME STANDARDS INPUT
Geo. J. Meyer Manufacturing Division of A-T-O, Inc.

IDENTIFICATION
PART NO. | OP | MDC
D4490300|10|TJ
1 2 3 4 5 6 7 8 9 0 1
1-11

LINE	PART NAME	MATERIAL	MO DAY YR	SET BY	RL	OPERATION DESCRIPTION	M B D C	FIXTURES, SPACERS	MC
010	SHAFT	C1141CF	041470	RJCB		FACE TURN GROOVE			A
020						FORM CHAMFER THREAD			A

LINE	SETUP DESCRIPTION	MASTER PART NO.	MC
030	LOAD PIECE INTO ROUND COLLET AND DOUBLE BUMP		A
040	INSIDE SPINDLE		

② PITCH DIAM. (55&56)
① PCT (31&33)

* "X" = SPINDLE INFEED (23)
* "T" = TIME VALUE
* "H" = HAND FEED
* "F" = SPEEDED FEED (21)

"M" = METHOD CONTINUATION
"1" = FREE FORMAT - SKIP 1
"2" = FREE FORMAT - SKIP 2
"S" = SIMO CUT
"N" = NOTES SUPPRESSED

MACHINE CUTTING DATA

LINE	SEQ NO.	CYCLE METHOD	CL	① SFM	SPEED	* ② FEED	CUT LENGTH	SE &A	BO	CU	TOOL AND HOLDER DESCRIPTION	*	MC
050	7	FC BOTH ENDS TO LGTH		260	610	00050	030006				2TBS KTFR 16C		A
060	8	CHAM BOTH ENDS HD FD		260	610 T02300						2SEN KSDN 16C		A
070	1	TN 1.0 DIA TO 1.015		160	610	00080	338 0006				TBS HERB ROLL TN		A
080	3	TN .75 DIA TO SIZE		175	890	00110	215 0006				TBS HERB ROLL TN		A
090	4	TN .562 DIA TO SIZE		130	890	00110	115 0006				TBS WS ROLL TN		A
100	8	CHAM .562 DIA HD FD		90	610 T00530						ISSUED ABOVE		A
110	8	CHAM .75 DIA HD FD		120	610 T00530						ISSUED ABOVE		A
120	9	GRV .125X.937 DIA		150	610 T01000						TG 44 GFT 64 B		A
130	10	GRV .187X.468 DIA		75	610 T00640						TG 46 KGT 64 B		A
140	7	FM .062R W 1.187 DIA		190	610 T02000						ISSUED IN SEQ 7		A
150	5	THRD TO SIZE		21	105 T01430						.75X10 VNC2A HSS	M	A
160											SHORT LEAD CHAS-	M	
170											ERS WITH DIE HD	M	
180		INSPECT-											
190		9 DIMENSIONS 1/10		USE SCALE								2	
200		4 TURNED DIAMETERS 1/10		USE 1 TO 2 MIC								2	
210		2 GROOVE DIAMETERS 1/10		USE KNIFE VERNIER								2	

Figure 3
THE METHOD ANALYSIS FOR AN OPERATION IS ENTERED ON THIS INPUT FORM.

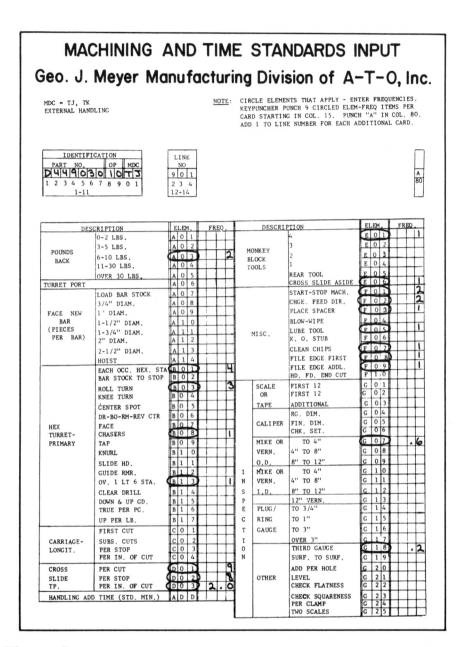

Figure 4
THE CHECKLIST IDENTIFYING MANUAL HANDLING MACRO ELEMENTS. APPLICABLE ELEMENTS ARE CIRCLED AND THEIR FREQUENCY NOTED.

MACHINE OPERATION SHEET REV LTR B PAGE 1 OF 1 (PAGE 1 OF 2)

PART NAME- SHAFT MAT- C1141CF PART NO- D449030 OP NO 10

OP DESC- FACE TURN GROOVE SET UP 3.883 PER PC .119 PCS/HR 8.4
FORM CHAMFER THREAD

MACH GR 311 MACH NO 593 SET BY- RJC DATE 04/14/70 MSTR PT NO

SET UP DESC- LOAD PIECE INTO ROUND COLLET AND DOUBLE BUMP
INSIDE SPINDLE

MOVE WORK TO & FROM MACHINE DURING CUT- NO BREAK SHARP CORNERS- NO

DURING CUT- NO COOLANT- NO FIX,SPCRS,ANG IRONS-

••••••••••••••••••••••••• MACHINE CUTTING DATA •••

SEQ NO	CYCLE METHOD	SFM	SPD RPM	FEED IPR	CLN	TL & HLDR DESC	CUT LNGTH	SCA	BO	NO CU	TOTAL TRV	SM /IN	SM /CUT	TOTAL TIME	LN NO
7	FC BOTH ENDS TO LGTH	260	610	0.0050	NO	TBS KTFR 16C	0.810	0.06	0.00	2	0.870	0.39	0.339	0.678	050
8	CHAM BOTH ENDS HD FD	260	610	0.0000	NO	SET KSDN 16C	0.000	0.00	0.00	2	0.000	0.00	0.000	0.230	060
1	TN 1.0 DIA TO 1.015	160	610	0.0080	NO	TBS HERB ROLL TN	3.380	0.06	0.00	1	3.440	0.24	0.826	0.826	070
3	TN .75 DIA TO SIZE	175	890	0.0110	NO	TBS HERB ROLL TN	2.150	0.06	0.00	1	2.210	0.12	0.265	0.265	080
4	TN .562 DIA TO SIZE	130	890	0.0110	NO	TBS WS ROLL TN	1.150	0.06	0.00	1	1.210	0.12	0.145	0.145	090
8	CHAM .562 DIA HD FD	90	610	0.0000	NO	ISSUED ABOVE	0.000	0.00	0.00	1	0.000	0.00	0.000	0.053	100
8	CHAM .75 DIA HD FD	120	610	0.0000	NO	ISSUED ABOVE	0.000	0.00	0.00	1	0.000	0.00	0.000	0.053	110
9	GRV .125X.937 DIA	150	610	0.0000	NO	TG 44 GFT 64 B	0.000	0.00	0.00	1	0.000	0.00	0.000	0.100	120
10	GRV .187X.468 DIA	75	610	0.0000	NO	TG 46 KGT 64 B	0.000	0.00	0.00	1	0.000	0.00	.00	0.364	130
7	FM .062R W 1.187 DIA	190	610	0.0000	NO	ISSUED IN SEQ 7	0.000	0.00	0.00	1	0.000	0.00	0.000	0.200	140
5	THRD TO SIZE	21	105	0.0000	NO	.75X10 VNC2A HSS	0.000	0.00	0.00	1	0.000	0.00	0.000	0.143	150
						SHORT LEAD CHAS-									160
						ERS WITH DIE HD									170

INSPECT- 180

9 DIMENSIONS 1/10 USE SCALE 190

4 TURNED DIAMETERS 1/10 USE 1 TO 2 MIC 200

2 GROOVE DIAMETERS 1/10 USE KNIFE VERNIER 210

TOTAL MACHINE TIME 2.757

Figure 5

THE OUTPUT REPORT CONSISTING OF ALL THE INFORMATION NEEDED FOR ENGINEERING BACKUP.

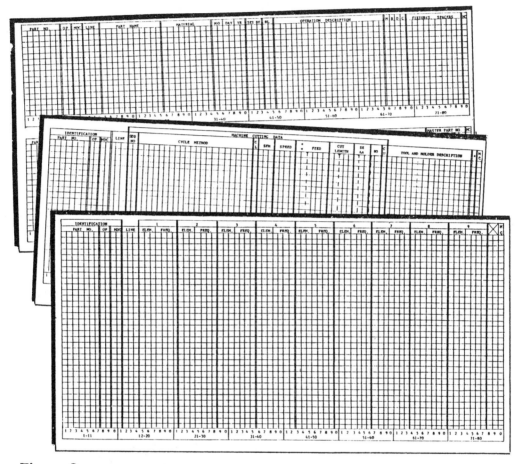

Figure 6
  THE VARIOUS INPUT FORMS FOR CREATING MAINTENANCE TRANSACTION CARDS.

Reprinted from Machine and Tool Blue Book, July, 1961.

Standard Time Data... for "guess-less" estimating

By E. A. Cyrol
E. A. Cyrol and Company
Management Consultants

■ Each new recession in the business cycle further proves that inefficiencies still exist in our manufacturing plants. When the profit squeeze gets tighter, further improvements in product design or in manufacturing methods are usually found. In our work as management consultants, we know from experience that a plant that is not using properly engineered work standards has labor costs that are at least 25% too high, and in many instances as high as 100% too high.

It is not uncommon for a person responsible for machine shop operations to challenge this remark with, "Well, I can see when people are working, that they are running the machines—they can't be just 50% efficient!" Well, it isn't difficult to find plants that look busy but still are producing at only 50%.

Using properly engineered work standards, incidentally, you are asking the workers to do profitable work during their eight-hour shift, and you do this by weeding out the wasted motions—the "looking" for tools—the trying of wrong set-ups.

Potential for improvement is large

Only 25% to 30% of the manhours in machine shop areas are presently based on standard time data; the rest of industry uses old-fashioned rate setting methods typified by their interest in, "How many pieces do we get an hour?"

It is quite obvious that there is a need for a real, systematic approach to cost reduction in machining areas through properly engineered work standards. Since the foundation of a cost reduction program is, first of all, a good basic work measurement technique, the standard data approach is the most practical for metalworking manufacturing.

What is Standard Time Data?

Simply stated, standard time data is a family of time values for the elements that make up an operation. Standard data analyzes the time values; time study records them. Put another way, rate settings are to standard data as mechanics are to engineering. Time studies are mechanical while standard data are creative operations.

Any work cycle that consists of elements that can be observed, described, timed and averaged can benefit from work standards. The time values are determined from time study observations of the common elements of an operation. The common elements of any turret lathe operation, for instance, would be "load piece in chuck—unload—index turret—etc."

The only sure way to establish standards for specific operations is to conduct quite a number, 75 to 100 as a rule, of time study observations of the operations that might be on milling machines, drill presses or turret lathes. If you were performing this many time studies, you would end up with 75 to 100 standards. Because standard data can be used over and over for various operations performed on one type of machine tool, you would end up with perhaps 2000 or more rate standards from 75 to 100 time studies.

Why Standard Time Data?

Many wonder, if both time studies and standard time data are used to determine how long it takes to perform an operation, why bother with the more complicated job of collecting standard time data?

Using the data involves analyzing elemental time values and assembling them in the most effective pattern. Practically all of the guess is taken out of estimating. Thus, time values are established that are applicable for many different jobs. Work standards established by individual time study of each job are a waste of time! Besides, they cost too much.

5 STEPS in Establishing Standard Time Data:

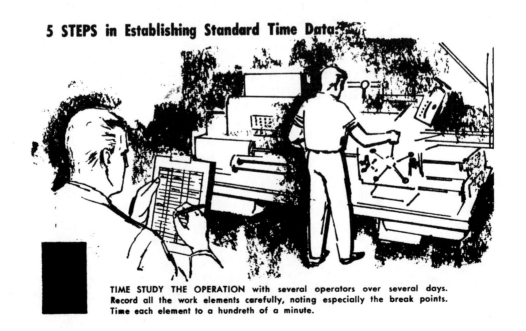

TIME STUDY THE OPERATION with several operators over several days. Record all the work elements carefully, noting especially the break points. Time each element to a hundreth of a minute.

Useful in selecting equipment

Suppose you are considering an improvement of one kind or another. It might be a new cutting tool, holding device, machine tool accessory, or something. Now, this improvement is going to cost you money. Suppose you are comparing a holding fixture, that uses wing nuts to hold the part down, to a hydraulic or air actuated clamping device. You would make a comparison, as to time, of the loading and unloading operation. After you have made the comparison and you have the data, you can go through all of your files and extend the production that would be applicable in this reduction to get the number of dollars per year that could be realized by the purchase of this equipment. This is a reference file that time study does not have!

One of the facts about standard time data development is that the average industrial engineer is unwilling to accept, at face value, that the time studies that he has in his files at the present time are probably completely useless as raw material for standard data! Many cost reduction programs may have been discarded before fruition because of this unfortunate fact. Time studies that are not taken according to a very clearly defined elemental breakdown pattern are useless for standard data because like elements cannot be compared to like elements.

One great problem in industrial plants is how to correctly reflect method improvements in new work standards! With standard data, the elements changed can have their time values easily up-dated. One of the main difficulties arising when changing standards after methods are changed is that the elements are not described in sufficient detail and the time values are not pinned down to the conditions sufficiently enough to permit continued improvements or continued revision.

The elements are described in a write-up. This seems to take a lot of time, but the development and the definition of standard data should be looked at the same way one looks at tooling up a job; a good, thorough job pays off many times over the long haul.

Five steps to standard data

When it is decided that time studies will be taken with the intention of using them in the determination of standard data, a few exploratory time studies must be taken of the operations for which the data are to be developed. These studies will permit the creation of a pattern that must then be followed. This pattern describes the element and indicates the point at which the stop watch must be read.

It is important that the watch must be read at precisely the same break point for each element; if this is not done, all succeeding steps will be useless.

After several time studies have been correctly taken, they are posted on sheets that are usually referred to as comparison sheets, spread sheets, or

The data collected must then be analyzed. Variables are charted. At this point look for possible improvements in methods and incorporate them in the standard work cycle. It is possible to realign elements to be able to assess the improvements on the work sheet. Standard data permits learning the best method possible by profiting from good element performance in many studies.

analysis sheets, so the information can be easily analyzed. Operational and set-up time values are similarly posted. About twenty-five studies should be posted before any attempt is made at analysis of the constants and variables, which can be easily spotted on the comparison sheets.

Testing the time data to determine the validity of the values is of utmost importance before preparing the standard data worksheet. If shortcomings show up, the data should be revised and improved at this time. For, a flaw ignored here will come up again and again to plague the man responsible for production standards.

It's good to take test studies of operations which weren't studied during the development of the original data, as well as the operations that were studied. Usually about ten studies taken according to the original elemental breakdown will test the standard data. Of course, these studies should be taken of large parts and small parts; operations having many and just a few elements should also be included. This is important so that the entire range of data is proven ahead of its "bath of fire" when it is released to the shop as production standards.

Finally, the standard data worksheet is designed. The layout is not important; but, it should be conducive to rapid calculation of the production standard. One such worksheet is used for each production standard calculated, and it is the record of the record of the standard determination. In writing up the study in permanent

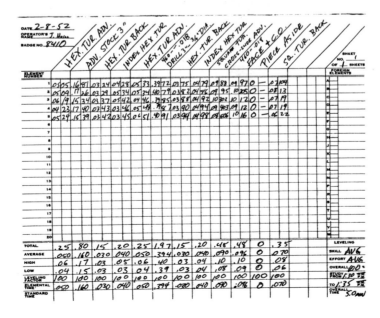

form include the detailed elemental descriptions, tables of time valves, etc.

How to use the worksheet

A production standard is calculated in this article to serve as a guide to the prescribed method of determining a production standard.

The use of properly developed standard time data eliminates the need for individual time studies of the operations that are to be placed on standard. Therefore, the standard data worksheets must be very carefully filled out with all the pertinent information regarding the operation. Should a grievance arise concerning a production standard, the worksheet and all its information, including descriptions and calculations, will be the basis of any investigation of the standard.

In the example, where gaging is required and gaging can be performed during a machine element, it is so designated on the worksheet.

If two simultaneous cuts are possible in an operation, the time allowed is the time for the longer cut.

When additional elements not included in the worksheet are found to be necessary, refer to the additional elements described in the write-up. If the desired element is included in this group, enter it on the worksheet in the "Additional Elements" box. If such an element cannot be found, a time study will be necessary to obtain this information. This new additional element should be recorded in the write-up for future reference and development of standard data for this element.

POST THE TIME STUDY DATA to comparison sheets where it can be easily analyzed. In this form, work elements that are constant and those that vary can be easily spotted.

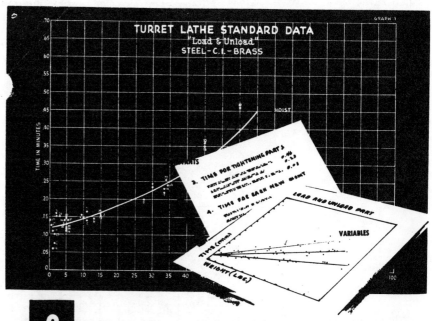

ANALYZE THE DATA to determine constant and variable elements. Chart the variables. At this point, look for possible methods improvements. Incorporate improvements in the standard work cycle.

PREFORM A CHECK STUDY to determine the validity of values found in previous step. Establish time values for the new elements resulting from methods changes.

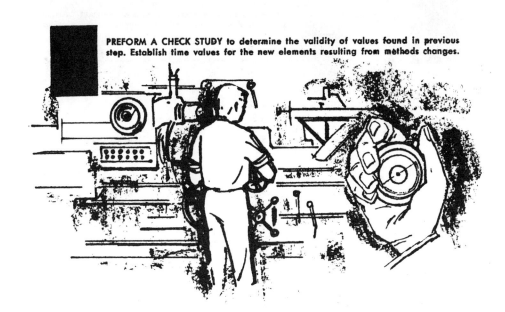

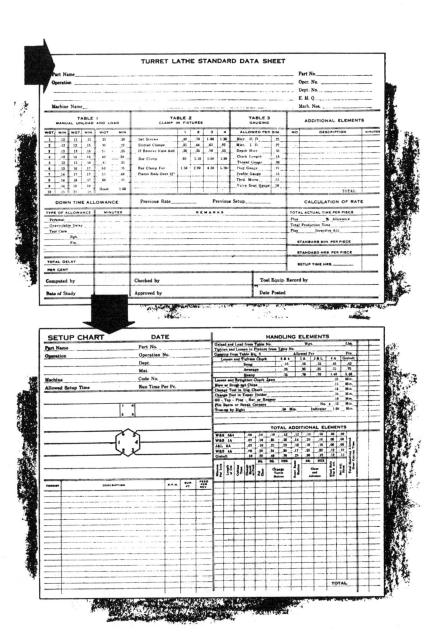

WRITE UP THE STUDY in permanent form, including detailed element descriptions, tables of time values, etc. Prepare standard data worksheets for finding total time for an operation.

5. **LOOSEN and RETIGHTEN CHUCK JAWS**
   When a part is apt to be sprung in average or heavy chucking, the chuck is loosened and retightened lightly before the finish cuts are taken. This applies to parts with thin flanges or hubs that have close tolerances that may be outside allowable limits if the part is sprung.

6. **BLOW OUT CHIPS**
   This element occurs in loading fixtures, soft jaws, and in checking diameters or blind holes. The time allowed covers the time required to pick up an air hose or brush, to clean the chips from the fixture or piece, and to replace the air hose or brush in its original position.

7. **CHANGE TOOL IN A SLIP CHUCK**
   This element includes the time required by the operator to remove a tool from the slip chuck, to dispose of the tool, to select a new tool, and to place the new tool into the slip chuck in preparation for the next operation.

8. **CHANGE TOOL IN A TAPER HOLDER**
   This element includes the time required by the operator to remove a tool from a taper tool holder, to dispose of the tool, to select a new tool, and to place the new tool into the taper tool holder in preparation for the next operation.
   An example of this would be changing a drill in a sleeve.

9. **OIL—TAP-PLUG-BAR-REAMER**
   This element covers the time required to apply oil with a brush to a tap, to a plug, to a steady rest bar, or to a reamer.

10. **FILE BURRS OR BREAK CORNERS**
    This element covers the time required to pick up a file, or scraper, to file burrs or break corners, and to replace the file or scraper to an out-of-the-way position. This time value is per corner.

11. **TRUE-UP**
    This element includes the time required by the operator to true-up the piece in the machine before finish tightening of the chuck. The element was found to vary according to the method used, i. e., true-up by sight, or true-up by dial indicator. The time values shown for each method are selected averages of the leveled observed occurrences. In the case of the dial indicator method, this includes also besides the actual true-up element, the setting up and removing of the dial indicator plus cleaning a spot to place the indicator fixture.

MACHINE MANIPULATION
   Machine manipulation times were found to be proportional to the size of the machine. For convenience in applying standard data, the

STANDARD TIME DATA EXAMPLE

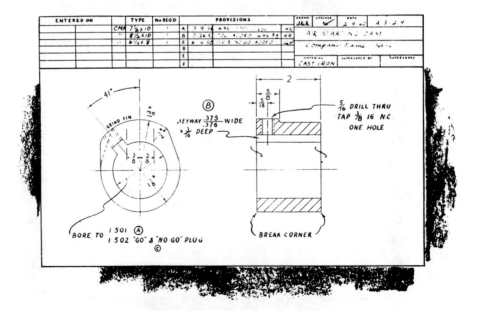

This is a sample part for which a production standard is to be determined through standard data.

STANDARD TIME DATA EXAMPLE

TURRET LATHE STANDARD DATA SHEET

Part Name: **Air Starting Cam** Part No. **T-7772**
Operation: **Drill · Face Bore** Oper. No. **10**
  Dept. No. **40**
  E.M.Q. **150**
Machine Name: **J & L 8A** Mach. Nos. **21611-21612**

TABLE 1 — MANUAL UNLOAD AND LOAD

WGT.	MIN.	WGT.	MIN.	WGT.	MIN.
1	.12	11	.15	25	.20
2	.12	12	.15	30	.23
3	(.13)	13	.16	35	.25
4	.13	14	.16	40	.28
5	.13	15	.16	45	.31
6	.13	16	.17	50	.35
7	.14	17	.17	55	.40
8	.14	18	.17	60	.45
9	.14	19	.18	Hoist	1.60
10	.15	20	.18		

TABLE 2 — CLAMP IN FIXTURES

	1	2	3	4
Set Screws	.40	.70	1.00	1.30
Slotted Clamps	.25	.44	.63	.82
If Remove Nuts Add	.20	.35	.50	.65
Bar Clamp	.60	1.10	1.50	1.90
Bar Clamp For Piston Rods Over 1½"	1.50	2.90	4.30	5.70

TABLE 3 — GAUGING

ALLOWED PER DIM.	
Micr. O.D.	.21
Micr. I.D.	(.27)
Depth Micr.	.35
Check Length 1/100	(.15)
Thread Gauge	.30
Plug Gauge	.15
Profile Gauge	.15
Thrd. Micro.	.51
Valve Seat Gauge	.50
	.27

ADDITIONAL ELEMENTS

NO.	DESCRIPTION	MINUTES
	None	
	TOTAL	X

Previous Rate **.15** Previous Setup **.5**

DOWN TIME ALLOWANCE

TYPE OF ALLOWANCE	MINUTES
Personal	20
Unavoidable Delay	15
Tool Care	
Rgh.	20
Fin.	
TOTAL DELAY	**55**
PER CENT	13.0%

REMARKS: **None**

CALCULATION OF RATE

Total Actual Time Per Piece	3.35
Plus 13.0 % Allowance	.44
Total Production Time	3.79
Plus 20 % Incentive Adj.	.76
STANDARD MIN. PER PIECE	**4.55**
STANDARD HRS. PER PIECE	.076
SETUP TIME HRS.	0.85

Computed by **WKC** Checked by **E.A.C.** Tool Equip. Record by **WKC**
Date of Study **2-14-59** Approved by **E.A.C.** Date Posted **2-14-59**

The front side of a Standard Data Sheet for turret lathes. On this side the part and operation are described; some of the manual elements are listed; allowance figures shown; and the standard data is calculated.

SETUP CHART

Part Name: **Cam** Part No. **T-7772**
Operation: **Drill - Face & Finish Bore** Operation No. **10**
  Dept. **40**
  Mat. **C-1**
Machine: **J & L 8A** Code No. **A-8**
Allowed Setup Time **0.85 HRS.** Run Time Per Pc. **.076**

FACE

Center Drill → Bore
Drill 1 13/32

HANDLING ELEMENTS

Unload and Load from Table No.		Wgt. 3 Lbs.		13
Tighten and Loosen in Fixture from Table No.				
Gauging from Table No. 3	Allowed Per		Pcs.	27

Loosen and Tighten Chuck	3 & 4	1A	J & L 4A	Gisholt	
Snug	.15	.15	.15	.45	.45
Average	.25	.35	(.35)	.75	.75	35
Heavy	.35	.70	.70	1.40	1.40

Loosen and Retighten Chuck Jaws	.25 Min.	
Blow or Brush out Chips	.15 Min.	
Change Tool in Slip Chuck	✓ .15 Min.	15
Change Tool in Taper Holder	.34 Min.	
Oil - Tap - Plug - Bar or Reamer	.10 Min.	
File Burrs or Break Corners DURING BORE (1 No. x .12 Min.)		(12)
True-up by Sight	.50 Min. Indicator 1.00 Min.	
GAUGE DURING DRILL HOLE TIME		

TOTAL ADDITIONAL ELEMENTS

W&S 3&4	.06	.10	.19	.12	.12	.16	.10	.06	.08
W&S 1A	.07	.10	.22	.14	.20	.16	.08	.08	.08
J&L 8A *	.07	.10	.21	.20	.12	.16	.18	.08	.08
W&S 4A	.08	.10	.26	.34	.17	.22	.20	.10	.10
Gisholt	.10	.20	.40	.35	.25	.30	.27	.12	.12

TURRET	DESCRIPTION	R.P.M.	SUR. FT	FEED PER REV.	Minutes Per Inch	Length of Cut	Cutting Time	Change Speed or Feed	SQ. Full Clear	SQ. Change Turret Station	HEX	SQ. Reset Same Station	SQ. Clear and Advance	HEX Blind Hole Tap & ect.	Set Adj. Head	Total Handling Allowed Over Cutting Time	
Hex 1	Center Drill	340	—	H.F	.42	3/8	.16	.07	—	—	.20	—	—	—	—	.27	43
Hex 2	Drill 1 13/32	243	90	.011	.37	2 3/8	.09	.07	—	—	.20	—	—	—	—	.27	1.16
Sq 1	Face End	340	235	.011	.27	3/4	.20	.07	—	—	—	—	.16	—	—	.25	43
Hex 4	Bore 1.501/1.502	489	193	.011	.19	2 1/4	.43	.07	—	—	.20	—	—	—	—	.27	70

				TOTAL	3.35

The back side of a Standard Data Sheet for turret lathes. On this side the setup is shown; the cutting elements are listed and detailed; machine times are calculated; machine handling times are listed; and cutting and machine times are summarized.

STANDARD TIME DATA EXAMPLE

SETUP CHART			
Part Name CAM		DATE	
		Part No. T-7772	
Operation DRILL-FACE & FINISH BORE		Operation No. 10	
		Dept. 40	
		Mat. C.I.	
Machine J&L 8A		Code No. A-B	
Allowed Setup Time 0.85 HRS		Run Time Per Pc. .076 HRS	

FACE

| | 1 | 4 | |
| | 2 | 3 | |

(hexagon diagram with positions 1-6, showing CENTER DRILL, DRILL 13/32, BORE)

TURRET	DESCRIPTION	R.P.M.	SUR. FT.	FEED PER REV.
HEX 1	CENTER DRILL	340		H.F.
HEX 2	DRILL 13/32	243	90	.011
SQ. 1	FACE END	340	235	.011
HEX 4	BORE 1.501	11.602 484	193	.011

TOOL SETUP CARD

Part Name AIR STARTING CAM Part No. T-7772

Operation No. 10 Description DRILL-FACE-BORE

Jig or Fixture Code No. 1   Name of Fixture 2-JAW CHUCK

TYPE OF TOOL	SIZE	CODE NO.
Drills	(1) CENTER DRILL	H.S.S.
	(1) 1 13/32" DRILL	H.S.S.
Reamers		
Taps		
Spot Facers		
Boring Tools	(1) BORE BAR + TOOL	CARB
Turning or Facing Tools	(1) FACE TOOL	CARB
Special Tools		

Remarks: SPECIAL JAWS #61330
+ 2-JAW CHUCK

Front side of a setup and tool card (left) is a carbon of the left back side of the Standard Data Sheet. Sent to the job with the tools, it tells the machine setup by giving tool positions, and speeds, feeds, number and sequence of the cuts. Back side of the card (right) identifies the part and operation, and lists the required tools. With this card, the crib attendance prepares the tools and fixtures in a kit and sends them to the job with the card.

TURRET LATHE STANDARD DATA
SET-UP STANDARD

THE FOLLOWING ELEMENTS ARE USED FOR DETERMINING THE SET-UP FOR ALL MACHINES:

MACHINE NOS. _21611 - 21612_

PART NO. _T-7772_   OPER. NO. _10_

1. Punch in and out		3.00
2. Get Tools & Mics from Crib		3.50
3. Clean Machine		1.00
4. Get & Store Tools ...... per tool .70 X— _4_		_2.80_
5. FIXTURES small hand lift		6.00
approximately (0-2 in. x 0-18 in.) hoist lift		12.00
approximately (2-4 in. x 15-30 in.) hoist lift		16.00

	#4, 1A & J&L	#4A & Gisholt	
CHUCK on and off (assume 3-jaw always on mach.)	(9.00)	13.00	_9.00_
STD JAW change (assume Std jaws are always on chuck)	7.00	10.00	
SOFT JAW change	7.00	10.00	
FACEPLATE	20.00	xx xx	
6. Get Ring, Etc., to Bore Jaws	3.50	4.00	
BORE JAWS (Minutes per inch) 3.30	all machines		
FACE JAWS (Minutes per inch) 4.00	all machines		
7. Tools (Load to Machine & Sizing)			
DRILL	(3.00)	3.00	_3.00_
REAMER	4.50	4.50	
TAP	3.20	3.20	
LIVE CENTER - _CENTER DRILL_	(2.00)	2.50	_2.00_
FORM .... first dimension	5.00	6.00	
Sizing each added tool use	3.00	3.50	
BORE .... Scale Dimension	4.00	5.00	
Micrometer Dimension	(5.00)	6.00	_5.00_
Sizing each added scale Dimension	2.00	2.50	
Sizing each added micrometer Dimension	3.00	3.50	
BEVEL .... Scale Dimension	2.50	3.50	
Micrometer Dimension	3.50	4.50	
Sizing each added scale Dimension	1.50	2.00	
Sizing each added micrometer Dimension	2.50	3.00	
TURN .... Scale Dimension	4.50	5.00	
Micrometer Dimension	6.00	6.50	
Sizing each added scale Dimension	2.50	3.00	
Sizing each added micrometer Dimension	4.00	4.50	
		SUB TOTAL:	_29.30_

This is the front side of a typical Turret Lathe Standard Data Setup sheet.

                                          SUB TOTAL: *29.30*
 FACE .... Scale Dimension (3.50) 4.00 *3.50*
        Micrometer Dimension 5.00 6.00
    Sizing each added scale Dimension 2.00 2.00
  Sizing each added micrometer Dimension 3.50 4.00
 DIEHEAD (Includes setting of stop) 11.00 11.00
ANY TOOL USING A BAR BEVEL
 HEAD .... add. 2.00 2.00
8. Set Stops ⊥ LONG Square Turret each (1.40) 1.60 *1.40*
     ⊥ CROSS Square Turret each (.30) .30 *.30*
       HEX Turret each (1.70) 2.00 *1.70*
  Screw Type Stop on Chuck Face each 5.00 5.00
   Ordinary Chuck Stop each 1.00 1.00

9. Line up a bar with pilot hole 1.30
10. Water connection on and off (steel jobs only) 4.00
11. Get and assemble inside micrometers or depth gauge .90
12. Get and assemble indicator gauge .... To Hex/Ways. 4.20
         To Square Turret 2.60
          Indicate O.D. 2.50
          Indicate I.D. 4.00
          Indicate Face 2.00

13. Inspection
 Travel time-punch card-check parallel, square, or
  concentricity .... (2.00) *2.00*
         Check O.D. .50 X ....
         Check I.D. .80 X *1* *.80*
        Check Length .40 X *1* *.40*
        Check Depth .40 X ....
 Check face and bore with arbor 5.00
             TOTAL *39.40*
 20 Min. Personal, 15 Min. Normal Delay Allowance 7.9% *3.11*
         TOTAL PRODUCTION TIME *42.51*
         Plus 20% Incentive Adjustment *8.50*
(MINUTES)   TOTAL INCENTIVE SET-UP STANDARD *51.01*
(HOURS)   TOTAL INCENTIVE SET-UP STANDARD *.85*
COMPUTED BY *W.K.C.*
DATE *2-14-59*

This is the back side of a typical Turret Lathe Standard Data Setup sheet showing the calculations for the setup standard.

CHAPTER 2

COMPUTER AIDED ESTIMATING

Machine job shop prepares quotations by computer

Automated manufacturing planning system is the basis for a system that generates quotations for one-third of the manual cost

In a job shop manufacturing firm, the price of a product is usually established by the firm through a quotation which, hopefully, will be accepted, and will also afford an acceptable profit margin to the firm. The preparation of a quotation is a crucial activity. Using the product specifications and drawing, an accurate and complete prediction of manufacturing costs must be determined. A quotation that is too high results in lost orders and opportunities for profits, and a quotation that is too low is likely to attract unprofitable business. The successful firm is the one that is able to avoid each of these extremes.

The preparation of the quotation is complex and time consuming. Generally it requires the services of the firm's most technically trained and experienced personnel whose services are not inexpensive in today's market. Furthermore, the majority of quotations do not result in an order.

These circumstances, for the most part, describe the situation confronted by a manufacturer of a variety of machined parts, produced to customer specifications. This company received an average of 400 inquiries for price quotations each month. Despite a competent and experienced estimating department, up to 450 requests for quotations were backlogged at any one time, and there was up to 25 days delay in processing the requests.

Management recognized that profitability and growth depended to a large extent upon responding effectively to requests for prices from potential customers. Confronted with the difficulty of securing and training additional competent staff in the estimating department, the president of the firm formed a team consisting of the general superintendent, the manager of manufacturing, and the author, as a consultant, to explore the feasibility of a computerized system of preparing price quotations.

The process of preparing a quotation, though highly complex and time consuming, fell within the capabilities of the computer. In addition, a computerized system offered a possibility of substantial savings in the cost estimating activity. The adopted system was dubbed AMPS — Automated Manufacturing Planning System. The preparation of a price quotation is, in essence, the development of the complete plan of manufacturing the product. It follows that, if the quotation results in an order, the process determined at quoting time becomes the shop routing.

Essentially, the computerized system follows the manual scheme of preparing quotations. This activity was broken down into three major segments:

1. Process routing — determining the type and sequence of operations necessary to produce the product;
2. Machine selection — determining the machine center on which each operation may be best accomplished;
3. Cost estimating — determining the process times and translating these times into cost by application of appropriate machine-hour rates.

The input to the system is a part description, along with a substantial amount of detailed specifications that serve to describe the part and its characteristics. The primary objective in organizing the data collection forms was to reduce the input needs to a format that could be supplied by an individual whose qualifications and experience may be considerably less than those needed to prepare quotations manually. This objective was accomplished by a series of input sheets, each calling for specific information, such as diameters, lengths, tolerances, hardness requirements, etc.

The package of input sheets required for a quotation consist in part of the following basic types: identifying information; general information and program options; material and order quantity specifications; end specifications; comments; diameter specifica-

Lloyd D. Doney
Marquette University
Milwaukee, Wisconsin

Lloyd D. Doney is Associate Professor of Industrial Management at Marquette University. He holds a BS degree from Michigan Technological University and MBA and PhD degrees from Louisiana State University. Dr. Doney has served as a research assistant at LSU and assistant professor at Bowling Green State University.

```
INQUIRY NO    DASH NO    CUSTOMER       CUST NO   DESCRIPTION   PART NUMBER   REV   DATE   PLANNER   CUST REF NO
   10                                      G        NO HT        BAR STOCK

INPUT DIAGNOSTICS

 ERR NO      LOCATION   DESCRIPTION

   83      INPUT CARD 3   AISI MATERIAL TYPE IS UNSPECIFIED.ERROR IN MATERIAL COST OR HEAT TREAT OPER.MAY OCCUR
   74      INPUT CARD 3   INSUFFICIENT DATA TO DETERMINE MATERIAL COST
                          ****CHECK DATA****
   26      INT DIA  1     INT DIA STOCK SPECIFIED.INVALID FOR BARSTOCK
   26      INT DIA  2     INT DIA STOCK SPECIFIED.INVALID FOR BARSTOCK
   26      INT DIA  4     INT DIA STOCK SPECIFIED.INVALID FOR BARSTOCK
   33      INT SHLD 4     WEB RADIUS AT OD UNSPECIFIED.ASSIGNED
                          RADIUS=.250
                 *************** NO FATAL INPUT ERRORS HAVE BEEN DETECTED ***************
                 *************** AMPS THANKS YOU FOR YOUR ERROR FREE WORK ***************
                 *************** WE SHALL NOW PROCEED TO PROCESS YOUR JOB ***************
```

Figure 1. Typical error messages produced by the input editor.

tions; external extras data; and holes data. The input sheets become a hard copy of the data presented to the system. They are used to obtain keypunched cards which are used as input to the computer.

INPUT ERRORS DETECTED

A significant feature of the system is an input diagnostic feature. It is a series of five FORTRAN IV subroutine subprograms that read in the input cards and examine the data supplied for omissions, obvious errors, and inconsistencies. This editing feature of the system contains over 90 error messages which describe the errors commonly made by those preparing the input. Certain of these errors are considered fatal because they terminate the processing of the job and preparation of the quotation. Other types of errors and omissions are considered nonfatal and generate warnings which are noted for the attention of the individual who later edits the quotation generated. In these cases, assumed values are inserted for items omitted in the input. However, the warnings note this fact and indicate what assumed values were used.

In addition to the editing for omissions, etc., the input editor can apply a generalized tolerance, stock allowances, and finish requirements to data which is supplied in its nominal form. As an example, the typical part drawing does not tolerance every dimension, but rather has a general machining tolerance and finish specification in the title block. Many of the dimensions are supplied as fractional. The AMPS input editor accepts these nominal sizes and automatically applies the prescribed general specifications. In this way, a minimum of input tolerancing and detail is required.

An example of error messages generated by the input editor is shown in Figure 1. Note that the errors are numbered and referenced in a manual for more comprehensive explanation as to the source of the input difficulty. The absence of fatal input errors enables the system to automatically proceed with the processing.

Another element of the input processing that is of special significance is that the system calculates both the finished and stock weight of the part. This was previously a manual activity that demanded a considerable amount of time and was subject to significant computational errors.

PROCESS AND MACHINES SELECTED

The total program consists of over 100 FORTRAN IV subroutines, each of which performs a particular function. The process routing segment consists of five subroutines, which are primarily logical routines, and cause the computer to select various operations, depending on product specifications. The number of operations needed to produce a part may vary from as few as 3 to as many as 15. The average is 8. The system produces a sequence number and a word description for each operation and, for certain operations, also generates detailed machining information. Finally, a subroutine for machine-group selection is called in from the process logic routine.

Like the process routing segment of the system, the machine-group selection subroutine subprograms — of which there are over 15 — are primarily logical routines which examine the input characteristics and make appropriate machine selections. The selections are based on such product specifications as size, tolerances, finish requirements, etc. Once the machine to be used for a particular operation is selected, the calculation of the setup and processing time can proceed.

The selected machine center, the operation and product specifications, stored in a common data area, provide the necessary input to determine setup and cycle times. The process times are generated by a series of different subroutine subprograms which are based on process and machine selections. These programs are primarily computational routines, though a certain amount of logic is often present to select the appropriate formula.

The computations are accomplished by a variety of methods including: tables based on product specifications; standard times based on MTM-generated standard data; and formulas, which were developed as a part of the project using multiple regression analysis. A particular subroutine may use one or all of these methods. Cycle times are stated on the quotation work sheet in terms of hours per piece.

The setup and cycle time provides input to the costing routine, which calculates the cost for that operation. If required, more than one method of costing may be employed at this point. In addition to determining the in-plant cost of the product, any outside costs associated with the item are included in a column headed "Pur." This includes raw materials, heat treating, outside machining operations, etc. Where vendors' pricing policies are known, the price lists are stored in an up-datable file and automatically accessed to compute the outside cost.

The costs for each of the operations are accumulated and totaled, and the price quotation is rendered at the conclusion of the processing of the job.

Figure 3 is an example of the output document generated by the system. Most of the column headings are self-explanatory, but certain items require additional comment:

• The 5-digit group designation is a numerical coding of the machine center. The first 2 digits indicate the department.

• Certain operations, such as shipping, receiving, inspection, etc., generate indirect costs which are included in the machine hour rates of the direct operations.

• MI rate and MI cost refer to the direct costing (marginal income) rate and cost system.

Other capabilities of the system which are significant include:

• The ability of the system to generate a quotation for up to six different inquiry quantities which may be supplied as an input item. In most instances, the size of an order has some impact upon the manner in which the order is processed. This, in turn, affects the cost.

• The print-back of selected input variables. In a computerized application of the complexity of AMPS, it is essential that appropriate input be provided to the system. In addition to the extensive editing capabilities, the system routinely prints back, for visual examination, selected, crucial input values on which the quotation is based.

The user may also select a print-back option which generates a more complete listing of input values. An example of this is shown in Figure 4.

• The system is currently operating on an IBM 360/50 OS system provided by a local computer service. Using a modular structure with overlays, the system requires approximately 290 K for execution.

• The system offers complete consistency. When new methods are implemented, the program is updated to include them. Afterward, each new job is quoted according to latest methods. There are no forgotten items as in the manual system.

• A list of tooling required is provided for each operation. This list is compared against the present tooling inventory so that the tooling costs can be accumulated for a particular quote.

At present, certain segments of the system require substantial verification. For this reason, the quotations generated are submitted to estimating personnel for editing. It is expected that this will be required for some time due

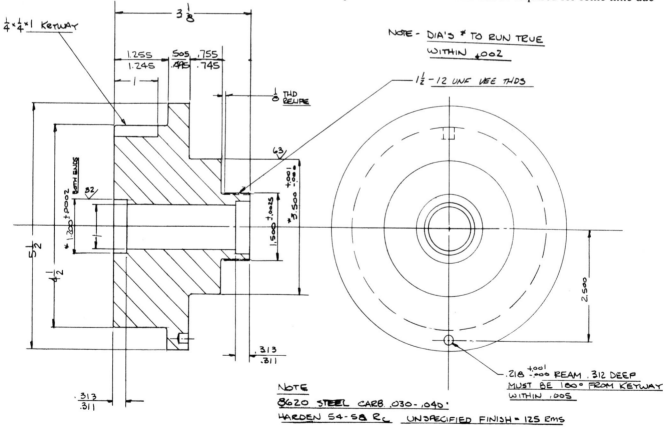

Figure 2. Customer's drawing of part which was used to generate price quotation shown in Figure 3.

```
  REF  2710              PART NO  XYZ                                        QUANTITY=   10                        AMPS
                                                      C                           EST   EST                  MI     MI
       OPER  GROUP  DESCRIPTION                    SETUP  CYCLE   HRS/PC    RATE  COST    PUR.      RATE   COST

        10   62500  MATERIAL   AISI   8617  $/CWT=18.90                                   5.19
                    GROSS COST WT.= 27.48LBS.

        20   15201  SAW AND SLUG HOLE               0.50   0.105   0.155   14.31   2.21            10.82   1.67
                    SAW  6.00 DIA -  3.235 LONG

        30   51601  RGH TURN HUB SIDE               0.99   0.152   0.251   13.93   3.49             8.68   2.18
                    END FACE 1  1.502 DIA.
                    IS FIRST ROUGHED SIDE
                    LEAVE .030 STOCK ON ALL SURFACES
                    ALTERNATE GROUP IS 52001

        40   51601  RGH TURN OTHER HUB SIDE         0.73   0.081   0.154   13.93   2.14             8.68   1.34
                    ALTERNATE GROUP IS 52001

        50   90000  75-02                                                                 0.50
                    WGT=  11.1  $/LB= 0.045  UNITS=    4.2

        60   51601  FIN TURN 1ST SIDE               1.74   0.224   0.398   13.93   5.55             8.68   3.46
                    END FACE 1  1.502 DIA.
                    IS FIRST TURNED SIDE
                    BORE IS FINISHED WITH
                    FIRST TURNED SIDE
                    PROCESS TOLER ON BORE IS 0.0050
                    ALTERNATE GROUP IS 52001

        70   51601  FIN TURN 2ND SIDE               1.42   0.114   0.256   13.93   3.56             8.68   2.22
                    ALTERNATE GROUP IS 52001

        80   31501  DRILL  1 HOLES ON 5.001/ 5.000 BC  0.80  0.028  0.108  12.95   1.40             7.88   0.85
                    USE .2031 DRILL  0.262 DEEP.CHF
                    FOR REAM.REAM 0.2190/0.2180 BY-
                    0.200 DEEP.
                    ALTERNATE GROUP IS 31502

        90   30301  MILL 0.2550/ 0.2450 KEYWAY IN   1.00   0.042   0.142   13.05   1.85             7.84   1.11
                    4.5100 DIA.NOTE .0050 ALIGNMT TO
                    DOWEL HOLE REMOVE BURRS

       100   37800  WASH,HANDLE CAREFULLY

       110   82100  VISUAL INSPECT

       120   90000  75- 8-15                                                              1.19
                    ALTERNATE GROUP IS 30305
                    WGT=   9.7  $/LB= 0.122  UNITS=   11.7

       130   44301  SURFACE GRIND BOTH SIDES        0.25   0.033   0.058   12.90   0.75             8.20   0.48
                    GRIND  4.5100 DIA. END FIRST
                    HOLD  3.1350 /  3.1150 OVERALL LGT
                    PIECES/TABLE LOAD =    10

       140   42005  GRIND 1.2052/ 1.2048 COUNTER BORE  2.00  0.102  0.302  13.02  3.94             8.24   2.49
                    BUMP CTRBORE FACE TO CLEAN UP
                    CHUCK O.D. IN HAND CHUCK,TRUE
                    UP BORE AND FACE
                    EST BASED ON .010STK IN BORE
                    EST BASED ON .010 FACE STOCK

       150   42005  GRIND 1.2052/ 1.2048 COUNTER BORE  2.00  0.102  0.302  13.02  3.94             8.24   2.49
                    BUMP CTRBORE FACE TO CLEAN UP
                    CHUCK O.D. IN HAND CHUCK,TRUE
                    UP BORE AND FACE
                    EST BASED ON .010STK IN BORE
                    EST BASED ON .010 FACE STOCK

       160   41001  GRIND 3.0060/ 3.0050 DIA        0.72   0.024   0.096   12.78   1.23             8.11   0.78
                    CYCLE TIME BASED ON .025STK ON DIA

       170   43601  GRIND 1.5000-12 VEETHREADS.TRI- 0.70   0.019   0.089   13.95   1.24             8.77   0.78
                    ANGLE SIZE 1.7737/ 1.7653

                    ****CHECK COATING SPECS****

       170   82100  FINAL INSP, ELECTRO MARK

       180   82200  SHIP PER HEADING

                                                     *** ESTIMATE COMPLETE ***
                                                                              ------   ------            ------
              TOTAL SETUP= $ 172.80  TOTAL CYCLE= $  14.02            TOTALS  $ 31.30   $ 6.88           $ 19.85
                                                 TOTAL ESTIMATED COST =$  38.19
              DIFF FACTOR = 2    MI FACTOR = .46  TOTAL MI PRICE = $  50.63
                                                             *
      *************************************************** END OF  2710 QUAN   10 *************************************
      *************************************************** END OF  2710 QUAN   10 *************************************
      *************************************************** END OF  2710 QUAN   10 *************************************
      *************************************************** END OF  2710 QUAN   10 *************************************
      *************************************************** END OF  2710 QUAN   10 *************************************
```

2207016

Figure 3. The quotation process: data from the original drawing was collected onto the input sheets. This was keypunched and fed into the computer, which generated a complete manufacturing plan and an estimated cost.

```
   INQUIRY NO  DASH NO      CUSTOMER      CUST NO  DESCRIPTION     PART NUMBER    REV     DATE     PLANNER    CUST REF NO
     2710                                           HUB             XYZ            01   11 30 70     TG

                                  *************** NO FATAL INPUT ERRORS HAVE BEEN DETECTED ***************
                                  *************** AMPS THANKS YOU FOR YOUR ERROR FREE WORK ***************
                                  *************** WE SHALL NOW PROCEED TC PROCESS YOUR JOB ***************



 ***************************************************** AMPS AUTO-LISTING *********************@*************************************
                                  ***********THIS ESTIMATE IS BASED ON FOLLOWING INFORMATICN***********

       QUANTITY=    0       MATL CLASS= 3       MATERIAL TYPE=H.R. BARSTCCK AISI NO= 8617   FAMILY NUMBER=24   TOOLING OPTION= 1
     FINISHED WT=   9.73  RGH STCCK WT=  25.91 MATL RATE CWT=  13.60   GENERAL TCL=0.0100    GENERAL FIN=125    THRU HDN = 20
   GEN STK ALLOW=0.3120    OVERALL LGT= 3.1350

   END 1 SPECS                WEB                                END 2 SPECS                WEB
    RMS= 63. TIR=.C050 STK=.312 CD= 0.0    ID= 0.0   DEPTH=0.0      RMS= 63. TIR=.0050 STK=.312 OD= 0.C   ID= 0.0   DEPTH=0.0

                 **  OUTSIDE DIAMETER SPECS  **                        ** ADJACENT SHOULDER  **
    NO MIN      MAX      RMS TIR     TAPER  MIN LGT MAX LGT STKCIA GROOVE ACJDIA RMS TIR     STK  RAD  LGT2    WEBCD   WEBID   WEBDPH   NC FIL
     1  1.4975  1.5025 1250.0075  0.0     0.6250  0.6350  6.000 0.0     3.000 1250.00180.0 0.0   0.0     0.0     0.0     0.0      0
     2  3.0050  3.0060  630.0020  0.0     0.7450  0.7550  6.000 0.0     5.500 1250.00260.0 0.030 C.0     0.0     0.0     0.0      0
     3  5.4950  5.5000 1250.0075  0.0     0.4900  0.5100  6.000 0.0     0.0   1250.00350.0 0.0   0.C     0.0     0.0     0.0      0
     4  4.5050  4.5100 1250.0075  0.C     1.2450  1.2550  6.000 0.0     5.500 1250.00320.0 0.030 0.0     0.0     0.0     0.0      0

                 **  INSIDE DIAMETER SPECS  **                        **  ADJACENT SHOULDER **
    NO MIN      MAX      RMS TIR     TAPER  MIN LGT MAX LGT STKID GRCOVE ADJDIA RMS   TIR    STK  RAD  LGT2    WEBOD   WEBID   WEBDPH   NO FIN
     1  1.2048  1.2052 32.0.0005  0.0     0.3110  0.3130  0.0    0.0     0.995125.0.00160.0 0.0   0.010   0.0     0.0     0.0      0
     2  0.9950  1.0050125.0.0075  C.0     2.4910  2.5110  0.0    0.0     0.0     0.0.0   0.0 0.0   0.0     0.0     0.0     0.0      0
     3  1.2048  1.2052 32.0.0005  0.0     0.3110  0.3130  0.0    0.0     0.995125.C.00160.0 0.0   0.010   0.0     0.0     0.0      0

                 **  BOLT CIRCLE HOLE SPECS  **                     *RADIAL* *SPEC-TOOLING*
    NO B.C.DIA NO MINSZE MAXSZE REAM TAPSZE TPI THRU DEPTH  DRILLSZE  DIA OIL  CTRSNK CTRBOR JIG MULTI    ALIGNMENT
     1  5.0010  1 0.2180 0.2190   1   0.0     0   0   0.200  0.2031     0  0     0      0    1    0

                 **  EXTERNAL  KEYWAY  SPECS **
    NO MAXWDH MINWDH DEPTH MINLGTH MAXLGTH QTY DEG DIA BLIND ENDMIL WOODRUF  ALIGNMENT   ALGNTOL
     1 0.2550 0.24500.2550  0.990   1.010   1   0   4   1     0      0      DOWEL HOLE  0.0050

                 **  EXTERNAL VEE-THRD SPECS **
    NO NOM.SZE THDS CLASS LENGTH DIA  SPEC PD TCL   MIN TRI MAX TRI RELF
   1 1.5000   12    2    0.500   1   1.44590.0083 1.7653  1.7737  1.

                        BARSTCCK CCST PER CWT INCLUDES QUANITY EXTRAS
```

Figure 4. Print-back of selected input variables permits visual examination of critical data.

to the complexity of the system. However, indications are that the AMPS quotation system cost is 30 percent of the manual method.

The estimating document, Figure 3, contains a substantial amount of operational detail attached to certain of the machining operations. Though this detail is not necessarily required for the preparation of the quotation, it serves to facilitate the auditing function by the estimators. Further, given the detail, the document may later serve as the basic shop order.

The fact that the system generates the basic shop document is particularly significant in that it serves as the starting point for all of the firm's activity that must result from the receipt of a particular order. Thus, the output of AMPS, when it attains the status of an order, provides the basic element of a total management information system that eventually could be expanded to encompass all activities of the firm. Documents generated by AMPS, for example, could serve as inputs to: a production scheduling and control system; a purchasing and inventory system; a financial accounting system; and a cost performance system.

The modular approach to the design of the system in terms of a series of FORTRAN IV subroutine subprograms affords a convenient arrangement for updating and maintaining the system. Machine rates, for example, which must change periodically, are contained in the cost subroutine. Thus, changes affect only this routine, causing no disruption of the other segments of the program. New machines, new processes, and other changes, necessitated by the dynamics of the industry, are capable of being introduced to the system similarly.

The present state of computerized quotations indicates that the firm's responses to requests for quotations can be reduced to a negligible amount of time. Under the present batch processing of jobs, quotes may be obtained in 24 hours or less from the service center. Ultimately, the firm hopes to obtain on-line capabilities to the service center's system.

The computer system also provides a consistency in process routing, machine selection, and arithmetic accuracy in cost computations that is not attainable in a human system of this complexity.

The experience described here suggests that a computerized system to generate quotations or bids presents an attractive alternative to manual methods presently being used in firms that obtain business through the submission of quotations or bids for eventual customer acceptance. Although the specifics of the system must be tailored to the products and the machine capabilities of a particular manufacturer or contractor, the general approach to the computerized preparation of quotations is one that has wide application in job shop industries.

A final comment concerning the role of top management in the development of successful data processing applications is appropriate. Most work on systems and their development points out that success in the development and installation of complex systems depends to a large extent on the leadership and participation of top management. In this instance, the foresight and the leadership provided by top management, as well as the cooperation received from key personnel in the industrial engineering department, were a vital ingredient in the success of the effort. ■

Donald G. Radke
Manager of Industrial Engineering
Engineering Products Division

E. R. Wagner Manufacturing Company, Inc.
4611 N. 32nd Street
Milwaukee, Wisconsin 53209

JOB SHOP COST ESTIMATING USING COMPUTER TECHNIQUES

E. R. Wagner is the nation's largest manufacturer of continuous hinges quoting on thousands of custom designed parts for hinges for cigarette lighters to hinges for collapsible houses to hinge floors to walls. The configuration for each customer part is unique within the limits of industry standard series. Material requirements, hole spacing, length, width, and shape are all variables engineered by customer requirements. The company structure places cost estimating under the control of Industrial Engineering who prepares detailed manufacturing process sheets to establish labor and material costs. The data base from which this information is drawn is the same multi-variable data used to establish incentive running standards and requires industrial engineering skills to utilize the base and establish costs.

Under this system, three industrial engineers spent 70-80 percent of their time preparing cost sheets and maintained a 2-4 week backlog of unfinished quotations. The net effect of this backlog was lost sales. Most of our inquires are for relatively simple parts and as such the customer expected a quick reply to his request, two weeks or less. With an average three week turn around many of our quotations were arriving after orders had been placed elsewhere. This in turn had an adverse effect on field sales personnel who became reluctant to accept "quick-quote" inquires. The needs of both sales and industrial engineering were combined into the following major goals:

1. Reduce industrial engineering time for estimating from 80 percent to 20 percent of total available hours.
2. Reduce the average time to complete an estimate from three weeks to four days.
3. Produce a program in which the majority of the estimating procedure can be performed by clerical personnel.
4. Maintain estimating accuracy to within 4 percent of actual cost (based on manufacturing cost after completion of orders).

In a brief experiment to increase the throughput of the department, the data base was condensed into a few major catagories with a single variable. This initially helped with the manufacturing cost calculation but still left the complex calculations for materials and packaging to the industrial engineer. The project was ended when it was discovered that estimating accuracy had slipped from a maximum error (estimated cost/actual cost) of 5 percent to nearly 20 percent. At this time, industrial engineering met with data processing to discuss the possibility of computerizing the data base.

An important requirement to initiating a computerized estimating system was to develop a data base which was compatible with the logic requirements of the computer. The first step was to back every individual calculation from the cost estimate. Each of these individual calculation steps was then analysed to determine from where each of the variables was derived. From this analysis, four basic sources of information were constructed.

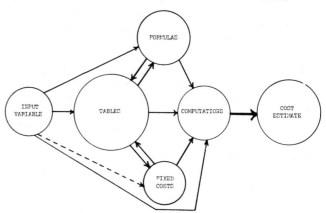

FIGURE 1 Information Sources Showing Their Relationship to the Computation and Final Cost Estimate

The first of these was labeled input variables. This information was classified as information requiring human decisions such as: the manufacturing steps as determined by process engineering, the manufacturing method as determined by industrial engineering, the part configuration per customer print, material specifications, and order quantities. The list was carefully reviewed to be quite certain that no item could be derived from some other source. The accuracy and advantages of the computer program necessitate a minimum of human decisions.

The second information source was a list of subroutine formulas whose answers were used as variables in the final cost computation, or as reference variables for searching tables, the third source of information.

Reference tables contain all the basic two level data. Included in this are manufacturing process standard data, packaging length/volume data, material handling data, die types, and hinge series data. The data in these tables may be referenced directly from input, from formula values, from other table values or any combination of these.

The last information source are the tables of fixed costs. These tables list base price information for materials, quantity extra adders, packaging material prices, running rates and set-up rates by machine, tooling prices, etc.

All of these information sources are utilized in the final computation section to determine the complete cost to manufacture. The figures are factored and summed to display an equalized cost per thousand for each order quantity requested.

The company's incentive system is an MTM based standard data program. The data was built in four major sections; materials, primary operations (essentially roll feed punch press operations), secondary operations (hand feed press operations) and packaging, each of which required calculations and a series of tables to determine costs. Wach section of data was divided into the four defined information fields.

The division of each section revealed several problems in the data construction. Subjective decisions were required to select table values which would have increased the amount of input information. To avoid this, major changes were made in data construction, additional tables built, and variable comparisons made in an effort to reduce or preferably to eliminate the additional input. Although these revisions greatly increased the size and complexity of the program it reduced substantially the knowledge and experience requirements of the estimator, a major goal in the program. Our success in reducing the input information requirements of the program were such that an 18 page booklet was produced containing all of the required input data.

After completing the division of each section into information groups a general procedure was written to facilitate programming the function. In this way a concise statement was available which defined the information sources, the calculation procedure and the resultant requirements.

To determine material requirements the computer is supplied with the running length of the part, its width, type of materials and the numerical equivalent of the style of hinge. The style number determines material added for hinge loops, material gauge, pin diameter, and scrap loss factor. The computer then calculates the cubic inches of wire and leaf material required, checks for the type of material of each component and converts both to weight. The weights are added together to yield the gross material content and the leaf material is reduced by the scrap loss factor to give the approximate finished weight of the part.

Each component weight is then extended by the customer quantity requirements to determine the total pounds of material required. The computer then searches the material pricing table for the $/CWT cost based on requirements and extends the total material value. When fixed material prices are not available from tables, such as with exotic metals, special customer requirements or materials with a highly voaltile price, the computer will signal the operator to provide a material price and extend values based on the supplied figure.

To determine the cost of the primary operation, the operator must key in additional information; running feed length, press RPM and the machine number. The computer then determines the setup hours, running standard hours and material handling allowances and extends the running and setup hours by their respective pricing rates. The standard running hours are a function of the number of press strokes per piece, the press RPM, the material handling allowances, and standard company allowances. All of the formulas and factors for labor standards are built into the program and valid incentive standards are produced with virtually 100 percent accuracy. The setup hours are stored in tables by machine number but special hours can be keyed in to cover non-standard conditions.

The computation of secondary operation costs is the second most complicated portion of the program. The format allows for up to five secondary operations which are calculated in a subroutine loop. Secondary costs are a summation of three major data groupings: part handling to and from the die, part handling to and from the press and part handling within the die (loading). The engineering data was first carefully screened to eliminate any reference to methods which could not be performed or which were beyond the range of specific equipment and other unlikely or impossible situations. This was done to reduce the size of the data base as much as possible. The data for part handling to and from the die had five machine class variables within seven length groupings within nine distinct methods of handling. Two separate sets of this data were built, one for hinges and another for separate hinge leaves.

The operator keys in a method code, the press machine class, the press number, and the "hinge" or "leaf" status of the part. Using this information plus the length and series data from the materials section the computer searches the tables for the proper value and places the figure in an array for summation.

To determine loading time the type of die to be used is specified by process engineering and keyed in by the operator. There are 18 types of dies possible, each yielding a set of loading times which vary with the length of the part. A total of 126 values are stored in a table and referenced for the loading value which is placed in the array with the other handling value for summation.

The part handling to and from the press is determined by a two letter code describing the part before and after the operation (plain or formed). The computer references virtually all the previously keyed in data to calculate and select the four values required to determine the allowance for this material handling portion of the operation standard. The part's length, width, gage, pin size, shape, weight, method of handling to the die, die loading, and handling out of the die are all factors in establishing the values. The four values summed for this portion are stocking up parts to the press, taking down parts from the press, the type of guarding used on the press, and the frequency of exchanging raw and finished parts containers. Each value is individually calculated and summed to yield a total value for this section. This total figure is placed into the array with the previous two values where they are added and extended by the pricing rate of the machine specified. The setup hours are also selected and extended by machine number. This sequence is repeated for each secondary operation.

The packaging section was designed to select the least expensive packaging method concurrent with customer requirements. The operator keys in the customers choice of either cartons or crates or if no requirements are known the computer is instructed to choose the least expensive container for the quantity of parts being quoted. The operator also supplies a coded digit relating to the nestability of the parts and whether the parts will be hand stacked or jumbled when shipped. Parts with hand fed secondary operations and parts over 24 inches long are generally stacked into containers.

With this information the computer selects both a carton and a crate which meet the volume/length configuration of the part, determines the number of parts per container within the weight limits of the container and the total parts being quoted, equalizes the number of parts per container for each lot quoted, determines the cost per part for the package and compares the two costs, selecting the lower. In this way each quantity being quoted is individually fitted to the most economical package. This selection is made from over 30 standard carton and crate sizes. The package costs for each quantity are then posted to an array for summation.

The summation of costs adds the material and packaging costs, which vary with quantity, to the standard running costs. The standard setup costs are divided by each quantity and the quotient added to the previous total. In this way an individual cost for up to four requested quantities is calculated simultaneously, displayed and printed out. It should be noted that all the input variables are entered at once; not with the calculation of each section as described.

Most problems encountered in developing and implementing the program were directly a product of program size. The number of variables, the size of the tables, the displayed outputs, stored versus supplied information; each item had to be looked at to determine its real need so that the program could be pared down to the limits of the available equipment. Programming errors, particularly truncated digits, led to frequent errors. Where five place numbers were at first thought to be sufficient, seven and eight place numbers are now used. Those combinations which were thought to be rare in occurance turned out to be commonplace. The rule ultimately followed was to build digit capacity to the extreme case to prevent errors caused by a dropped digit.

COST ESTIMATE INPUT SHEET

PROJECT NO. 09-666   CUSTOMER NAME SAMPLE #1

SERIES 765   BLANK LENGTH 042.000   ACTUAL LENGTH 042.000

WIDTH 03.875   HINGE MAT. CLASS 41   PIN MAT. CLASS 96

FEED 3.000   HINGE OR LEAF H   PIN DIAMETER ____   GAGE ____

PACKAGING-WEIGHT RESTRICTION R   CONDITION S   CONTAINER E

NESTING FACTOR 1

_____ENTER

MATERIAL PRICES   QQ1 H__.__ P__.__   QQ2 H__.__ P__.__
                  QQ3 H__.__ P__.__   QQ4 H__.__ P__.__

QUANTITY ___.500   1.000   5.000   15.000

SETUP CD BC WC 123101101   SETUP HOURS 8.500

PRIMARY OPER RPM 265   DESC RUN HINGE

SETUP CD BC WD 102105101   SETUP HOURS 3.500

OPER 2 DIE 05 CYCLE 09H1 METHOD PP DESC PERFORATE

SETUP CD BC WC 102105101   SETUP HOURS 1.200

OPER 3 DIE 05 CYCLE 09H1 METHOD PP DESC PERFORATE

SETUP CD BC WD 102101101   SETUP HOURS 3.200

OPER 4 DIE 23 CYCLE 06H4 METHOD PF DESC FORM

SETUP CD BE CW __/__/__ S/STD __.__   SETUP __.__

OPER 5 DIE __ CYCLE ____ METHOD __ DESC _____

SETUP DC BC WD __/__/__ S/STD __.__   SETUP __.__

OPER 6 DIE __ CYCLE ____ METHOD __ DESC _____

** TOOLING COSTS HOURS ___.__   MATERIAL __.__.__
                                                    PF 1
                                                    PA 1

FIGURE 2
COST ESTIMATE INPUT SHEET
SHOWING ALL THE INFORMATION REQUIRED TO
PROCESS THE COST ESTIMATE

To maintain the accuracy of the system the variable pricing information in storage is printed out and sent to the material procurement section for monthly review of prices. Standard packaging is stored and referenced by its part number and any change in dimensions or price is immediately inforporated throughout the system with the processing of engineering changes. Machine pricing rates were not built into the program, rather, the program references active accounting files to extend pricing costs. Whenever possible, where data was presently available it was not duplicated into the estimating program; the program was instructed to search the total company files to obtain the latest figures for calculations. In this way, only unnumbered raw materials are separately tabulated in the program itself.

The net effect of the program has been to reduce the industrial engineering time spent estimating from 100 hours per week to 34 hours per week while maintaining an estimating error record of only 3 percent. The estimating job itself has been simplified to the point where much of it is done by student help and the estimating backlog has decreased from weeks to slightly over 24 hours. Future plans include firm price quotations, given by sales personnel equipped with a computer terminal, in minutes for the phone-in customer.

The computer used is an IBM System 3 Model 15 with a Model 3275 CCP display unit and a 3284 printer. The program language is RPGII. The total core required for the program is 33,700 bites. Program execution time is 1 to 2½ minutes to enter the variable data, three seconds to transmit the information to the central processor, six seconds to process the data, and seven seconds to transmit and display the information at the terminal. The program required over 800 manhours to develop at a cost in excess of $12,000. The payback period was less than 0.5 years.

```
SER LG/BLK ACT/LG  WIDTH  HR FM *** ESTIMAT *** 09-666  CUST  SAMPLE #1
765 042000 042000   03875  41 96  FEED 3000  H OR L H  PIN DIA 1250  GAGE 057
WT RES R  COND  S  CONT  E  NEST 1         3 COIL MIN 000356 '000712 001068
MAT/CWT H 01425  P 01610  H 01425  P 01610  H 01363  P 01610  H 01285  P 01610
* QTY * 000500 $ 041752 001000 $ 041738 005000 $ 039922 015000 $ 037630
             $      SU HRS    CYCLE   METHOD   OPER DESC   STANDARD    COST/M
SETUP CDBCWC  1230101  08500                                  8.500    131.75
PRIMARY OPN  RPM 265                          RUN HINGE       2.107     49.93
SETUP CDBCWC  1020501  03500                                  3.500
OPN 2  DIE 05                    09H1   PP    PERFORATE       1.254     19.19
SETUP CDBCWC  1020501  01200                                  1.200
OPN 3  DIE 05                    09H1   PP    PERFORATE       1.254     19.19
SETUP CDBCWC  1020101  03200                                  3.200     33.60
OPN14  DIE 23                    06H4   PF    FORM            3.391     56.80
SETUP CDBCWC          00000             S/STD
OPN 5  DIE
SETUP CDBCWC          00000             S/STD
OPN 6  DIE
* TOOLING COST HRS         MAT           SU    165.35   RN    145.11
PK1     500   8289   PK2    500  8289   PK3   714  8289   PK4    750   8289
QUOTE QTY1 COST/M     936.83  SU    330.70  PKG    43.50   NWT/M   3059
QUOTE QTY2 COST/M     771.34  SU    165.35  PKG    43.50   LWT/M   2929
QUOTE QTY3 COST/M     607.86  SU     33.07  PKG    30.46   PWT/M    149
QUOTE QTY4 COST/M     561.51  SU     11.02  PKG    29.00   GWT/M   3078
```

FIGURE 3 The Completed Cost Estimate as Printed at the Computer Terminal

The Use Of Minicomputers In Cost Engineering

Raymond R. Wenig, Member AACE
President
Internation Management Services, Inc.
70 Boston Port Road
Wayland, MA 01778

Reprinted by special permission from the (1976) Transactions of the American Association of Cost Engineers. Copyright © (1976) by American Association of Cost Engineers, 308 Monongahela Building, Morgantown, WV 26505.

Minicomputers represent a significant technological phenomena that is happening at an avalanche pace. There is almost no management, technical service, or other functional area that has not been impacted by minicomputers in one way or another.

The purpose of this paper is to briefly review the capabilities of minicomputers and evaluate their use, application, and impact in cost engineering systems. Several guidelines and checklists will be introduced to help the user cope with minicomputers and obtain successful results from their capabilities..

MINICOMPUTER CAPABILITIES

There are two main things to remember about a minicomputer:

1. A minicomputer is a computer and requires the same data care, and careful management as that needed by other computers.
2. A minicomputer uses short binary words to perform all of its processing. Each word is usually 16 binary bits long which is equivalent to four (4) decimal digits or two (2) alphanumeric characters. To produce realistic computations or handle a meaningful string of information, the minicomputer must process <u>multiple</u> words of data.

Most operational aspects of a minicomputer are closely related to a large-scale computer except that it is smaller, slower, and less costly. Today's minicomputer takes from 2 to 60 cubic feet of space, requires minimal environmental interfaces, and costs from $500 to $500,000 with major stops at $5,000 and $50,000.

There are two major differences between minicomputers and large-scale computers. One difference is that the minicomputer is located near the user and not in a closed computer room. This proximity implies user responsibility and user control of the system and its utilization.

The second major difference is that a minicomputer operates on events or data values as they occur or are presented to the system rather than on collected batches of similar data as most large-scale computers operate. This event response capability means that the minicomputer is dedicated to responding to user requests or data as they are created. It is often referred to as real-time or interactive computing.

These two major differences of minicomputers over larger computers provide a significant capability that allows the minicomputers to perform exceptionally well in any system where a small group of users is located in close proximity to one another and where their data can be processed as it occurs. There are limits to a minicomputer's processing power so the complexity of the computing steps must be reasonable if the data is going to be effectively handled by a minicomputer. Usually, the minicomputer can handle any transaction oriented processing such as data logging, account processing, data base extract and updating, and other separable processing applications.

The net impact of all of these characteristics is that minicomputers provide capabilities that are ideally suited to many of the needs and requirements of cost engineering systems. The following sections will outline some of the major areas where minicomputers can support and aide cost engineers in the performance of their jobs.

DATA LOGGING

The cost engineer is usually concerned with the collection of data that accurately represents the occurrences and conditions of the environment that the engineer is responsible for monitoring and controlling. The minicomputer can be an effective tool for aiding in the process of logging, validating, and collecting cost data.

The logging of data involves the digitizing and storing of data values as near to their natural occurrence as possible. The recording can be done via direct reading instrumentation, through human operator recordings, or as a by-product of another required activity.

Minicomputers are natural devices for data logging. They can handle many different signal sources and can respond to the signals as they occur. These attributes are extremely useful for reducing the labor efforts of recording and for improving the accuracy of the data. The computational capabilities of the minicomputer provide extensive editing, testing, and validation to be performed on each data element and on the overall pattern or logical groups of data elements (such as a work shift or a selected time period).

The resulting logged cost or operational values will be stored by a minicomputer in a computer readable form and usually in a format and media that is easily accessed or processed to generate cost engineering reports and studies. The processing of the values can be done at the time of logging, if the computations are not too involved or time-consuming, or they can be performed at a later time on groups or batches of data. The processing itself can usually be done on the minicomputer or it can be done on other computers with the transfer of the logged data files.

The major problems in minicomputer based data logging are in the following areas:

1. Timing of data element recording. The frequency and rate of arrival of data values must be within the recording capabilities of the minicomputer.
2. All signals must be in digital form. Many operational values are in analog form (voltages, flows, etc.). They must be converted to pure numeric form (a value at a specific point in time rather than a fluctuating signal).
3. The numeric values must be within the range of the minicomputer's capabilities. Extremely small, very large, or widely variant numeric values can be taxing on the small computer word structure of a minicomputer. The best performance can be obtained on values that contain from 4 to 17 significant figures.

A. Examples of Minicomputer Data Logging

The following examples present logical area where minicomputers can and have been used to log and develop data files on costs and operations parameters:

1. Continuous Production Logging: A minicomputer can be

coupled directly to analog monitoring and control instrumentation on continuous processing systems through analog to digital converters. The converters will digitize major values of the process, and feed them to the minicomputer for editing, control computation, costing, and logging.

For example, the minicomputer might log the following values:
a. Raw material inputs.
b. Intermediate output levels.
c. Waste materials.
d. Output qualities.
e. Energy inputs.
f. Process times or flows.

From this logged data the minicomputer could calculate and report the following factors:
a. Time period output (hour, shift, etc.).
b. Cost of energy.
c. Efficiency of process.
d. Cost of product produced.
e. Shift variances.
f. Output and efficiency trends.

2. Discrete Production Logging: In the discrete environment the minicomputer will record action events as they occur. The system can generate source indentifiers, time and place of occurence, and other pertinent data parameters. The minicomputer is an ideal random event sensing unit. It can handle a reasonable number of event recording units and can validate the accuracy of the input. It can also update control records and compute various standard cost and time performance values.

An example of a discrete logging situation would be a jobshop production floor reporting system. Such a system will consist of recording terminals scattered throughout the production area connected to a recording minicomputer. The terminals could be equipped with badge readers, magnetic card readers, punched card readers, and/or direct entry keys, switches or other sensors. They will also contain a time clock generator, a storage bugger, and some transmission control circuitry to send the data to the host minicomputer. The terminals' output capabilities vary from simple lights, to one line displays, to full video screens to printers. Their major difference from the sensors used in the continuous environment is that they must be coupled to a human user and must be two-way communicators handling both inputs and some form of output.

The minicomputer based discrete logging system provides a wide range of interactive capabilities during the process of data logging. These capabilities make the minicomputer based system more powerful than the older hard-wired simple data recording systems. An example of these extended capabilities is provided in the following list:
a. Operator log in (usually badge insertion into terminal).
b. Validation of proper individual using terminal.

    c. Input of type of transaction (terminals can support a wide range of functions).
    d. Establish terminal controls to receive the indicated transaction type.
    e. Request and prompt the operator to enter appropriate values into terminal.
    f. Validate values as entered, inform operator of any errors and options for correction.
    g. Display results of entry for operator to review.
    h. Select, generate, and display the next logical step for the operator to perform.
    i. Transmit acceptable data to the minicomputer.

At the end of a logical data logging sequence, the minicomputer system can be programmed to perform the following steps:
    a. Update the proper work element file with the job data.
    b. Compute performance times, costs, work output values, and post to appropriate files.
    c. Generate the next operation in sequence for any completed work elements.
    d. Review overall shop floor load and generate a rebalance of assignments.

In addition, the system can respond to special logging inputs, such as:
    a. Equipment breakdowns.
    b. Defective work.
    c. Quality variances.
    d. Workers absent.
    e. Inconsistent work step reporting.
    f. Missing or extensively delayed entries.

B. <u>Benefits of Minicomputer Data Logging</u>

The use of minicomputers for data logging provides a more comprehensive analysis to be made of data values at their time of occurrence or creation. The programmable intelligence of a minicomputer can be used to:

    1. Validate.
    2. Analyze.
    3. Update.
    4. Logic test.
    5. Sequence test.
    6. Limit test.

Also, the programmable intelligence of a minicomputer can be used to condition the data values into a proper and logical format. The fact that all this is done within a tolerable duration of time and uses low-cost, dedicated minicomputers means that more local information is available on the events, and it is more accurate and usable for the next effort(s) in the environment.

ACTIVITY PROFILING

The process of activity profiling involves the monitoring or tracking of the trends of various operational values within an environment and detecting significant (or adverse) shifts, directions, or changes in the activity pattern or profile. Most activity profiling requires a continuous monitoring of selected parameters from the environment and the computation of trend lines or directions.

This type of processing is well suited to the minicomputer. It involves continuous processing of a repetitive nature and depends on the analysis of real-time events soon after they have occurred. The output of profile processing is usually a graphic display, a plotted output, some form of monitoring panel, or a simple continuous form printed report.

The cost engineering aspects of activity profiling can be found in two major areas; one area is the continuous dedicated monitoring of a fixed facility, and the other is the temporary monitoring of the profiles of a special situation. In the fixed situation, the minicomputer operates in a manner similar to the data logging situation; however, it also processes the data profiles (continuously, at set times in groups, or when it is not doing logging) and generates the appropriate profile outputs.

Temporary profiling with minicomputers is a new and interesting capability for the cost engineer. The small size of the minicomputer provides for easy transportability of the unit. The use of small file devices (such as cassette or cartridge magnetic tapes and/or floppy disk drives) provides good data logging capabilities with lightweight portability. The use of programmable (or easily modified) data logging instrumentation allows the profiling minicomputer system to be moved to the site of a component or element that the cost engineer wants to study. The minicomputer system can be connected to the equipment, program validation steps taken, and a profile of events can be recorded and then processed to provide the cost engineer with output for evaluating and/or tuning the process. The portable profiling system may be left in place for an extended period of time to check the results of the engineering adjustments.

A. <u>Examples of Minicomputers in Activity Profiling</u>

The use of activity profiling minicomputer systems is a relatively new and dynamic area. The flexibility of the minicomputer as well as its reliability and low cost are opening new areas for activity profiling on a continuous basis. The following paragraphs describe two applications in these areas:

    1. <u>Energy Management and Control Systems:</u>
        The effective use and control of energy (especially electrical energy) has become an urgent problem to which cost engineers can apply their skills. The temporary energy source cutoff, escalating prices, and declining supplies of current fuels have all pushed the concept of energy conservation and control from wasteful non-concern to the forefront of management interest at all levels in heavy energy using firms.

        The use of electrical energy in a commercial/industrial firm is an excellent example of an activity profiling situation. The costs of delivery and consumption of such energy varies with total amounts, and the levels of demand (loads) of the organization. The more consistent the load level of electrical power, the lower the charges for the power. As the load levels vary up or down, the power costs go up by significant factors.

        The use of minicomputers for energy management and control systems involves coupling them to all major energy use devices and monitoring the usage and demand profile of electrical energy. As certain devices increase their demand for power, the minicomputer will try to reduce power uses in other variable power uses (such as heating and ventilating systems) in order to maintain an even power load in the controlled system. Proven savings in power costs ranging from 20% to 50% have been attained through this use of minicomputer systems.

The minicomputer based energy management and control systems will usually perform data logging, activity profiling, and direct power use management. In addition, most of the systems can drive control monitors and displays and generate printed energy utilization and cost control reports. Some of the larger systems even have extra capacity and are used to perform facility security control functions.

2. Mass Transportation Vehicle Engineering:

The rebirth of the mass-transportation concept has brought sophisticated design engineering tools into play to design and build performance and cost effective vehicles for reliable and automated (minicomputer) control. The new breed of mass-transit vehicles (such as rail cars, subway cars, and electrical trams) are complex integrated systems. Their design and contruction involves strict adherence to performance specifications while at the same time maintaining strict cost controls.

The use of temporary minicomputer monitoring and activity profiling systems has become a major tool of the cost and design engineers in the transportation vehicle field. The minicomputer(s) can be installed in test vehicles and instrumented to various points in the system (the internal electronics of the system often provide readily available monitoring points). The vehicle can then be operated in a test mode or subjected to expected environmental actions and the results monitored, logged, and profiled via the minicomputer system.

The profile results are often directly output by the minicomputer system (on displays, printers, strip recorders, etc.). The cost engineers can quickly evaluate the impact of their designs, adjust them, and then rerun the tests to determine the next set of experimental results. Some minicomputer systems can keep track of large numbers of trial profiles and then compute the differences and pinpoint the most effective profile in the group.

B. Benefits of Minicomputer Activity Profiling

The minicomputer adds the capability of rapid output of results and its computational analysis power in addition to the attributes of data logging to provide a positive benefit for activity profiling. Its abilities to be programmed to detect and respond to adverse profiles, sound alarms, or even issue correction commands, provides benefits in the areas of timely response and controlled modifications that could not even be handled by manual methods.

In temporary profiling situations the minicomputer can provide quick and low cost evaluations of numerous parameters plus a locally controlled output and computational analysis capability that can be used as needed by the cost and design engineers.

COST HISTORIES

The cost engineer is partly a good historian. The occurrence of actions and events has a pattern that influences the costs of products and operations. The cost engineer must keep track of cost histories and reuse their content to establish control and/or target cost levels, learning curves, standards, and other parameters.

The collection and use of cost histories has two major problems. One problem is that costs occur at "local" points throughout an organization; the second is that cost details quickly mount up to an enormous volume of data.

The traditional way to handle cost histories has been to develop summaries of selected parameters and capture them as a by-product of other operations in a centralized computer system. This process has several drawbacks including:

1. Saving of only composite parameters.
2. Slow process of building up cost histories.
3. Difficulty in accessing the history data (batch processing at low priority on central computer systems).
4. Inability to adequately analyze cost relationships.

The minicomputer is starting to have an impact on improving the recording, maintenance, and use of cost history information. The major factors of this impact are found in their use in field locations close to the natural occurrence of the cost elements and the availability of low cost data storage facilities.

Minicomputers can quickly generate and store detailed cost histories on local operations. This can usually be done as a by-product of normal operations. The amount of detailed cost data to be stored is directly related to the size and volume of the local operation. It is usually possible to build and maintain minicomputer data files that consist of several million characters of data. This data can be stored on magnetic tape or disk and quickly processed to derive cost reports. The decreasing costs of magnetic storage media means that the data can be stored as archives for long periods of time at reasonable cost and safety.

Another major factor in the use of minicomputers for cost history keeping is that then can also be used (depending on programming capabilities) to access and retrieve cost history data and to process comparative computational analyses. These analyses can be processed locally and on request by the resident cost engineer.

With the improving minicomputer data base programming languages, it is becoming easier for the non-computer professional to set up, process, and report information in meaningful ways on the minicomputer. The use of simple interactive languages like BASIC can be very easy for cost engineers. It can give them flexible processing capabilities using local cost files and the local minicomputer.

On a periodic basis, the local minicomputer cost files can be summarized and transmitted to a higher level organizational unit. This will allow automated transfer of pre-validated cost data suitable for integration purposes while still retaining details at local levels.

A. <u>Example of Minicomputers in Cost History Generation</u>

A good example of minicomputers for cost history recording is their use in field construction sites. The local minicomputer is used to maintain full data collection and processing at the site. The processing includes:

1. Worker control.

2. Task analysis.
  3. Material accounting.
  4. Time monitoring.
  5. Vendor/subcontractor accounting.
  6. Active job control.

The results of all data logging builds a composite set of project cost history and performance files. These files can be accessed on the site minicomputer by the project cost engineer(s) to review the cost histories, including such things as:

  1. Rates of expenditure.
  2. Expenditures versus budgets.
  3. Expenditures versus progress.
  4. Labor rate variances.
  5. Other analyses.

The availability of an on-site minicomputer will give the local personnel access, control and responsibility for their use and should increase their interest, concern, and utilization of the cost data elements. In addition, they can extract summary project data from the system for submittal (or transmission) to the central office for consolidation into company reports.

B. Benefits of Cost Histories on Minicomputers

The primary benefits are found in local control and use of the data. The cost histories will be more accurate and will be more utilized to aid in reviewing and improving performance. The rapid access and retrieval of current cost information as well as history trends from a minicomputer will give the cost engineer a new tool to work with. It should allow him to track cost trends more closely and detect and respond to variances while then can be compensated for and not adversely affect the work efforts.

COST ESTIMATING

Cost engineers spend a significant amount of their time in estimating the expected costs of new items and in setting plans, limits, and controls on the costs' expectations. Most of cost estimating is based on analysis of relevant past data tempered with judgmental expectations on the work item(s) to be performed or produced and the conditions under which they will be produced.

The cost estimating process involves access and review of prior cost histories and the propositional evaluation of current expectations. This work requires retrieval, processing, extrapolating, combining, modifying, re-projecting, comparing, and finally choosing an expected cost pattern. Many of the activities of the cost estimation process involve data retrieval or computation. Both of these efforts as well as most of the above steps can be helped with effective computer processing.

To be an effective aid in cost estimating, the computer processing must be:

  1. Available on demand.
  2. Responsive to user requests.
  3. Capable of rapid generation and display of results.
  4. Easy to change data and rerun projections.
  5. Easy to give command instructions.
  6. Able to access history files.

A minicomputer system can effectively meet these requirements. In addition, it can be economical and locally available to the estimator.

A. Examples of Cost Estimating Using Minicomputers

One of the most advanced applications of minicomputers in cost estimation is found in the design of electronic circuit boards (many of which are used to build minicomputers). In this application the designer uses an interactive minicomputer to develop the specifications of the circuit board. The minicomputer then selects components from a data file, looks up their current prices, does a volume level calculation, factors in rejections, test time and other components, and extrapolates an expected cost estimate.

Another example is found in the interactive graphics applications on minicomputers for architectural specification generation of standardized structures such as schools, hospitals, etc. Some systems are able to support the generation of structural size specifications and then generate a reasonable cost estimating for producing the end result.

B. Benefits of Cost Estimating Using Minicomputers

The major benefits of utilizing minicomputers are found in time savings, improved accuracy, more experimentation, and more consistent standards for the estimating process. Much still depends on the data files and on the user. The minicomputer provides an available data access and manipulation tool through which the cost engineer can look at more aspects of the cost estimate and hopefully apply more of his professional expertise to produce a better cost estimate.

DISTRIBUTED COST MANAGEMENT

Cost engineering and cost management are functions that are dynamically dispersed throughout any organization. Costs are abstracted valuations of happenings in the real world. These happenings are individual occurrences that take place continuously throughout the environment. Most cost engineering systems try to sample these individual occurrences, record them, and combine them into a total collage with other happenings. This compression almost always leads to inaccuracies, inconsistencies, and loss of credibility.

Distributed cost management uses minicomputers as local processing support for the cost engineering activities. They can be used to log data, generate profiles, keep histories, develop estimates, and activity assist in keeping the operational performance under reasonable control. All of these efforts can be performed in the local environments near where the cost elements occur.

The minicomputer provides dependable, low cost data processing power that can be used in support of local cost engineering activities. It provides interactive processing support for local problems using local data. It also can supply all or part of its data to other computing systems in an immediately usable and pre-validated form. This means that most centrally needed cost data can be supplied or generated from the local minicomputer files.

Distributed cost management is the coming method of the future. It blends the support of computer processing with the logic power of the human user and transfers the resultant appropriate for sharing values to other organizational levels. The minicomputer is making possible the operation of distributed cost management at the departmental or project levels. Soon microcomputers will bring the distributed level down to the work station and personal levels.

A. Example of Distributed Cost Management Using Minicomputers

A distributed cost management system depends on the collection, processing, and use of data to local levels with the passage or sharing of appropriate information with other systems. A good example can be found in a multiple department manufacturing plant where each functional unit has a local minicomputer to support its data collection and processing operations. As work items move to another department, the minicomputers can select and pass along appropriate data and associated control information. If a department needs to know status information on a component in another unit, it can make a request for access to the holding unit's computer and receive the results through its own local minicomputer.

B. Benefits of Distributed Cost Management Using Minicomputers

The distributed cost management concept allows local control to be maintained over cost and operational data elements. It also reduces the cost of data creation by sharing all elements for as long as they are valid. In addition, local users will have more immediate access to their cost data and to pertinent data from other sources. Upper levels of the organization will be supported by regular or special request cost summary reports that are generated from the local data bases.

THE CHALLENGE OF MINICOMPUTER SYSTEMS

The above sections have briefly outlined some areas of minicomputer use in the field of cost engineering. The successful use of minicomputers requires the review and control of several potential problem areas. The major difficulties of today lie in the following areas:

1. Too many minicomputer vendors offering too many products.
2. Too little end user support.
3. Programming can be expensive.
4. Maintenance is slow in arriving when needed.
5. Users must develop good data disciplines.
6. The systems are always changing.
7. Data security can be a problem.

These problems can be solved if the prospective user recognizes their existence and takes proper management steps to eliminate them or reduce their impact. The explosive growth of minicomputers has left their slower growing support efforts severely strapped and usually inadequate for helping the ultimate user.

The maturing status of the minicomputer as an effective data processing tool will help to diminish the above problems. In addition, the logic of using minicomputers for cost engineering applications such as those outlined above will help cost engineers and managers to assume the pioneering risks of such systems, and reap the rewards of using them to help enhance their professional performance.

Cost Estimating and Production Planning at the Preliminary Design Stage

By Daniel E. Strohecker
Project Manager
The Battelle Development Corporation

At the preliminary design stage, a number of product-related areas must be considered before a decision can be made as to the profitability of the item. Cost estimating and production planning are two critical areas that should be examined. Generally, there are only limited data available at the preliminary design stage. When a request for quotation is received on a new product, normally, there is limited time available to make a cost analysis or determine the manufacturing sequence to be used.

Most companies rely on historical cost information to prepare cost data on new products. This technique works well whem similar products have been produced that generated historical data. If such historical data is not available, new-product estimates can be grossly inaccurate when based on this technique.

Most companies desire to have a better feel for the cost of manufacturing items at the preliminary design stage. However, it has been recognized that to improve estimating at this stage requires a considerable amount of effort. At least part of the effort must be applied to detailed manufacturing plans before starting the estimate. Obtainment of the best estimate requires knowledge of the means for manufacturing the item as well as the cost of all materials going into the product.

PROGRAM CHARACTERISTICS

In 1967, the U. S. Army Missile Command contracted with Battelle Memorial Institute to develop computerized cost estimating procedures for the inert components of solid-propellant rocket motor cases. The objective of the computer program was to estimate the fabricating and materials cost for the inert components of any given rocket motor case. The program was to encompass all major components of a solid-propellant rocket motor except the propellant and assembly or loading of propellant. The program was not to consider development costs or R & D costs. This effort was assumed to be complete at the start of manufacturing.

Typical components included in the program were the motor case, its skirts, and its closures. Certain other limitations were also placed on the program. The program was to estimate costs for a motor case with a maximum diameter of 60 inches and a maximum length of 180 inches. The materials of construction to be considered were those commonly used in

Copyright Society of
Manufacturing Engineers 1970

For presentation at its
Engineering Conferences

rocket motor cases, such as high-strength steels, aluminum, and titanium. The program developed is flexible enough to include other materials if their characteristics of fabrication are similar to one of those listed in the program.

Since minor components such as hangars, wiring brackets, mechanical-test coupons, and major documentation efforts were neglected, the cost estimate calculated by this program was expected to be only within 10 to 20 percent of the actual cost. The program was to produce cost estimates of sufficient accuracy for a designer to use in his early design exercises. He could use the estimates for making comparisons between various optimum-performance rocket designs and determine which design could be produced at minimum cost.

PROGRAM USE

To use the computer program, it is necessary to have certain input information on the characteristics of the motor case. The major dimensions, general configuration, number of components, size and number of important openings, slots, etc., are needed. In addition, materials, costs, labor costs, and tolerances placed on the important dimensions must be specified.

The output of the program is in terms of dollars per item at any given quantity level. In addition to the total costs which are estimated for the finished motor case, the output also shows costs broken down by component and fabrication method. Consequently, the program can be used to examine fabrication costs for methods other than the lowest cost method. This is how the planning phase was entered into the program. The program considers several different methods for fabricating each component and selects the least-cost method for each particular component. The costs for all feasible fabricating methods are shown for each component. This additional information on alternative fabrication methods is useful in planning future production. For instance, even though one method of manufacture may entail lower cost than another, a particular plant may have idle capacity to manufacture by a slightly higher cost method that in the long run might entail lower cost to them. Also, a make or buy decision can be made based on an examination of the costs for a given component.

The program is quite simple to operate. It is written entirely in FORTRAN IV for a IBM 7094-2 computer. It runs as a standard FORTRAN job with card input and printer output. The source deck consists of roughly 6600 cards, nearly half of which are explanatory comment cards. Normal run time for the program is less than 1 minute per estimate.

FLOW CHART

The computer flow chart presented in Figure 1 is very similar to the flow of material (for a rocket motor case) through a typical fabrication shop. The flow chart shows the fabricating processes considered (for each component) and the costs that are calculated, compared, and combined to form the final estimates. The calculated costs may require the determination of up to four equations for each component and method of manufacture. Some of the equations obtain considerable complexity since they were derived to cover all possible variations in manufacturing procedure.

The computer program considers a rocket motor consisting of up to five components. The components are designated as: 1. Case, 2. Head (Fore), 3. Head (Aft), 4. Skirt (Fore) 5. Skirt (Aft). It was recognized of course that all of these components might not be required. Consequently, the last three components were considered optional and the designer specifies whether these are to be included.

PROGRAM DEVELOPMENT

Before the computer program could be written, it was necessary to determine what component characteristics affected the cost of manufacture and the interrelationships. A number of component characteristics were selected and classified into geometric parameters, process parameters, and other parameters. In general, all the physical characteristics of the component were considered as geometric parameters. The process parameters were more concerned with the conditions of the process that would affect cost. Other parameters were those which did not fit in either of the other two categories. These included quantity, labor rates, material costs, etc. This group of parameters should be handled as input data. The program user must determine the values of the input data.

PROGRAM APPLICATION

A number of special input forms were designed to simplify the use of the program. An explanation was placed on each of the input sheets so that the user could readily determine how he should fill out each sheet. It has been found in practice that an engineer can review the data sheets and information on the program initially within about 4 hours. The input sheets can normally be completed within 1 hour. These sheets are then used directly to prepare the keypunched cards. The cards are fed directly into the computer. Consequently, the total time required for a given analysis can be quite short.

Initially, when the program was being made operative, a number of cost estimates were hand calculated on sample motor cases. It was found that to do a complete analysis required approximately 1 man-month of hand-calculating time. Comparison of the times required by the two methods, 1 man-month as opposed to approximately 6 engineering hours, indicates the time and cost savings that can be obtained by utilization of the computer technique.

PROGRAM OPERATION

The program as shown in the flow chart of Figure 1 begins by reading the input data and calculating the components weights. The component weights are frequently used in the subsequent cost calculations. This is because weight was found to be a very sensitive parameter in some of the fabricating techniques. It also directly affects the cost of moving material from one production station to another. The program then proceeds to estimate the forming and machining cost for each component. After this, the cost of joining the components is estimated and the costs of such auxiliary final operations as heat treating, moving the material between stations, and testing the components are summarized.

As mentioned previously, several possible methods of fabricating each component are considered and compared in the program. There are five different methods for forming the case, which include extrusion, roll bending, roll forging, making the case from drawn tubing, or hydrostatically extruding tubing. The costs for each of these forming methods is estimated separately and then compared. The computer selects the minimum component cost and thereafter ignores the others. However, all five of the costs are printed out so that the user can determine how these other costs or methods of manufacturing would compare with the least-cost method. This practice is followed for costing each of the components.

There are also several alternatives built into the program. These are marked as Alternative 1 and Alternative 2.

Alternative I considers that the case and head (fore) are formed separately and then joined together, while Alternative II considers that both components are made at one time as a single part. The program has a decision function which determines whether Alternative II is feasible and, if so, whether it carries along the separate cost of this technique as well as Alternative I.

EXAMPLE OF PROGRAM USE

The computer cost analysis of alpha pressure-bottle shows the program potential for commercial hardware. The parameters and dimensions of this bottle are shown in Figure 2. The bottle consists of three components and is to be made of AM-350 stainless steel. A quantity of 10,000 bottles are to be made in a facility with labor rates including all burden of $6.50/hr for unskilled labor and $10.50/hr for skilled labor. The first printout from the computer shows the input data. Consequently, this can be readily checked with that desired to determine where there were any preliminary errors. The input data used in this estimate are shown in Figures 3 and 4. Here the number of components required as well as the quantity to be produced and the labor rates used in the calculations are given. Next, the component tables are printed for each of the components. The various parameters marked there show the material used, material costs, the physical dimensions of the part to be made, the shape to be used, tolerance, and whether the parts are to be welded or not.

The input data for the holes table and the milling table on the various components are also given. Here the position of the hole relative to the center line of the part is indicated, the amount of material that must be drilled through, and the number of identical holes of a given diameter. If the holes are to be tapped, the number of threads per inch is indicated. Also, if nut plates are to be installed, it is indicated in this particular table of information. In a similar manner, the number of impressions for the milling table at a given location are indicated. Also, the length, width, and depth of the milling cut is indicated. The holes table and milling table for a given component are not printed unless the user has specified that there will be holes and/or impressions in the component.

The cost estimates calculated by the program for the alpha bottle are printed out in a sequence similar to that followed by the flow chart, as shown in Figures 5 through 10.

The estimates are given for each individual component first. The column headings across the page are the names of the various forming processes that can be used to make a given component. There are also a number of subcosts entries listed on the left side of each table. The subcosts are given for each manufacturing step. The total costs are shown at the bottom of each page. A final line is then given which indicates the lowest cost method of manufacturing the component and what this cost is. Also, when the program determines from the input data that a particular method of manufacture is not feasible, this fact is indicated by the printing out the words "not feasible", in the corresponding column heading. That particular column would then contain only zero cost entries.

A subcost is not necessarily relevant to all forming costs and a zero entry for a subcost does not necessarily mean that this is a calculated zero. There are a number of possibilities in this case: (1) the forming method itself may not be feasible, (2) the subcost is not pertinent to that particular forming method, and (3) it is not pertinent for the particular motor case being estimated.

Following the cost summaries for each component in Figures 5 through 8, a final estimated cost sheet for joining the components is given in Figure 9. A total estimated cost of the finished pressure bottle which summarizes the various fabrication methods selected is combined into the final summary cost sheet shown in Figure 10.

In this program a broad-brush approach has been given to the planning function. It would be quite feasible, however, to break down each individual cost in terms of every processing step normally specified on a planning sheet. It would be possible to have direct readouts from a computer program, giving all the necessary planning steps as well as the cost for each of these steps. The computer could also select the lowest cost manufacturing procedure. Taking the planning one step further, it would be possible to predict loading on individual machines for a given time in the future. Consequently, the capacity to produce a given component as well as its cost would be known.

The time-saving feature of the computer program used for cost-estimating has been shown. Once the planning function is added to the cost-estimating function, the total time saving possible using such programs is very significant. When the detailed cost-estimating alone can be reduced from 1 man-month to approximately 4 hours, the initial saving involved in the planning function can be seen to be quite sizable. It is believed that the man power required to perform the planning function in large plants could be reduced as much as 95 percent by the introduction of the concepts developed in this program. Considerable effort would be required to obtain this goal initially. However, the savings to be obtained through introduction of such a system in a normal manufacturing sequence make it quite promising at this point. The program performed here on an example pressure bottle illustrates a technique which can be used with excellent results.

When the costs calculated in this program were compared with actual costs, a correlation of ± 10 percent was obtained. This is considerably better than the normal military hardware-production contracts, which can vary up to 100 percent. With this technique, contractors no longer need to depend on a gross procedure of calculating costs. The use and expansion of the procedures outlined in this paper are dependent only upon imagination and future desires of manufacturing management.

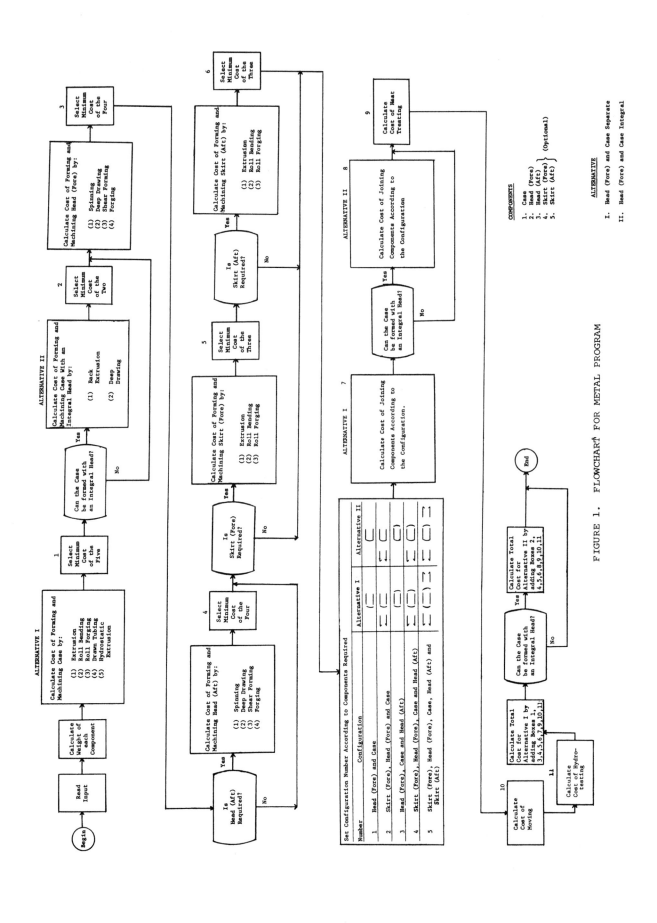

FIGURE 1. FLOWCHART FOR METAL PROGRAM

97

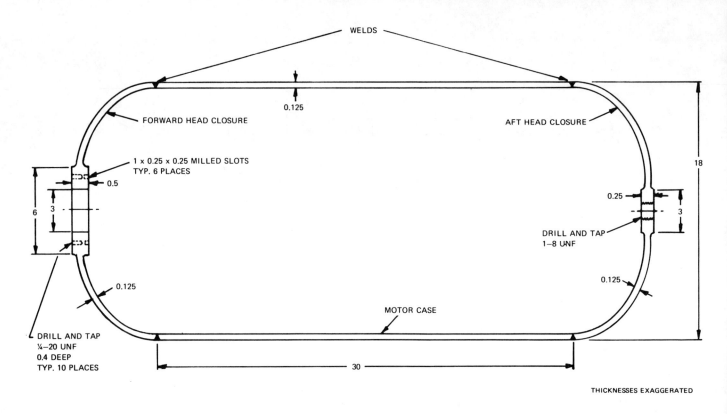

FIGURE 2. DIMENSIONS OF ALPHA PRESSURE BOTTLE

ROCKET MOTOR CASE COST ESTIMATION
METAL PROGRAM

ALPHA PRESSURE BOTTLE

COMPONENTS REQUIRED

CASE
HEAD(FORE)
HEAD(AFT)

UNSKILLED LABOR RATE (L1) = 6.50
SKILLED LABOR RATE (L2) = 10.50
QUANTITY PRODUCED (Q) = 10000

COMPONENT TABLE FOR CASE		COMPONENT TABLE FOR HEAD(FORE)		COMPONENT TABLE FOR HEAD(AFT)	
AM350S.STEEL	MAT	AM350S.STEEL	MAT	AM350S.STEEL	MAT
1.25	M1	1.25	M1	1.25	M1
.90	M2	.90	M2	.90	M2
30.000	H	18.000	D	18.000	D
18.000	D	6.000	D1	3.000	D1
17.750	D1	3.000	DD	1.000	DD
.125	T	.500	T	.250	T
.125	T2	.125	T2	.125	T2
.005	TOL	.005	TOL	.005	TOL
C	SHAPE	ELL	SHAPE	ELL	SHAPE
YES	LWI	MOD	CONTOUR	SIM	CONTOUR

FIGURE 3. COMPUTER PRINTOUT OF INPUT DATA FOR
ALPHA PRESSURE BOTTLE

HOLE TABLE

HEAD(FORE)

DAP	T3	N	DD	TPI	P	NPI
1	.500	1	3.000	NO	0.000	NO
1	.400	10	.250	YES	20.000	NO

HEAD(AFT)

DAP	T3	N	DD	TPI	P	NPI
1	.250	1	1.000	YES	8.000	NO

MILLING TABLE

HEAD(FORE)

W	L	T3	N
.250	1.000	.250	6

WELDING INFORMATION TABLE FOR CASE

2.000	MWF
NO	EBI

FIGURE 4. COMPUTER PRINTOUT OF INPUT DATA FOR ALPHA PRESSURE BOTTLE

(ALTERNATIVE I - CASE AND HEAD(FORE) SEPARATE)

COST BREAKDOWN FOR
FORMING AND MACHINING
CASE

FORMING METHOD

		EXTRUSION	ROLL BENDING	ROLL FORGING	DRAWN TUBING	HYDROSTATIC EXTRUSION
		NOT FEASIBLE			NOT FEASIBLE	NOT FEASIBLE
	MATERIAL	0.00	75.28	58.66	0.00	0.00
	MATERIAL PREP.	0.00	427.83	427.83	0.00	0.00
	TOOLING	0.00	0.00	0.00	0.00	0.00
	FORMING	0.00	2.98	149.45	0.00	0.00
	JOINING	0.00	57.21	0.00	0.00	0.00
SUB-	SHEAR FORMING	0.00	0.00	0.00	0.00	0.00
COSTS	EXPANDING	0.00	1.77	0.00	0.00	0.00
	HOLE MACH.	0.00	0.00	0.00	0.00	0.00
	MILL MACH.	0.00	0.00	0.00	0.00	0.00
	FLANGE MACH.	0.00	15.82	15.82	0.00	0.00
	TOTAL	0.00	580.89	651.76	0.00	0.00

FORMING METHOD SELECTED (ON BASIS OF MINIMUM COST) - ROLL BENDING

COST -   580.89

FIGURE 5.   COMPUTER OUTPUT DATA FOR ALPHA PRESSURE BOTTLE

(ALTERNATIVE II - CASE AND HEAD(FORE) INTEGRAL)

COST BREAKDOWN FOR
FORMING AND MACHINING
CASE WITH INTEGRAL HEAD(FORE)

FORMING METHOD

		BACK EXTRUSION	DEEP DRAWING
		NOT FEASIBLE	
	MATERIAL	0.00	143.52
	MATERIAL PREP.	0.00	1.00
	TOOLING	0.00	1.51
	FORMING	0.00	13.94
SUB-	HOLE MACH. CASE	0.00	0.00
COSTS	MILL MACH. CASE	0.00	0.00
	FLANGE MACH. CASE	0.00	15.82
	HOLE MACH. HEAD(FORE)	0.00	11.10
	MILL MACH. HEAD(FORE)	0.00	25.31
	FLANGE MACH. HEAD(FORE)	0.00	15.82
	TOTAL	0.00	228.02

FORMING METHOD SELECTED (ON BASIS OF MINIMUM COST) - DEEP DRAWING

COST -   228.02

FIGURE 6.   COMPUTER OUTPUT DATA FOR ALPHA PRESSURE BOTTLE

COST BREAKDOWN FOR
FORMING AND MACHINING
HEAD (FORE)

		FORMING METHOD			
		SPINNING	DEEP DRAWING	SHEAR FORMING	FORGING
	MATERIAL	91.44	92.30	86.26	135.15
	MATERIAL PREP.	8.35	8.35	5.39	83.90
	TOOLING	0.00	.38	.36	1.92
SUB-	FORMING	16.11	7.17	142.55	187.93
COSTS	CONTOUR MACH.	22.71	22.71	0.00	20.15
	HOLE MACH.	11.10	11.10	11.10	11.10
	MILL MACH.	25.31	25.31	25.31	25.31
	FLANGE MACH.	15.82	15.82	15.82	15.82
	TOTAL	190.84	183.14	286.79	481.28

FORMING METHOD SELECTED (ON BASIS OF MINIMUM COST) - DEEP DRAWING

COST -   183.14

FIGURE 7.   COMPUTER OUTPUT DATA FOR ALPHA PRESSURE BOTTLE

COST BREAKDOWN FOR
FORMING AND MACHINING
HEAD (AFT)

		FORMING METHOD			
		SPINNING	DEEP DRAWING	SHEAR FORMING	FORGING
	MATERIAL	26.74	27.09	43.13	115.74
	MATERIAL PREP.	8.35	8.35	5.33	83.90
	TOOLING	0.00	.38	.36	1.92
SUB-	FORMING	8.58	5.91	98.66	187.93
COSTS	CONTOUR MACH.	12.45	12.45	0.00	20.15
	HOLE MACH.	1.04	1.04	1.04	1.04
	MILL MACH.	0.00	0.00	0.00	0.00
	FLANGE MACH.	15.82	15.82	15.82	15.82
	TOTAL	72.98	71.04	164.34	426.50

FORMING METHOD SELECTED (ON BASIS OF MINIMUM COST) - DEEP DRAWING

COST -   71.04

FIGURE 8.   COMPUTER OUTPUT DATA FOR ALPHA PRESSURE BOTTLE

COST BREAKDOWN FOR
JOINING COMPONENTS

ALTERNATIVE I
(CASE AND HEAD(FORE) SEPARATE)
PLUS HEAD(AFT)

SUB-COSTS

    JOINING HEAD(FORE), HEAD(AFT)
        TO CASE   143.49

TOTAL   143.49

ALTERNATIVE II
(CASE AND HEAD(FORE) INTEGRAL)
PLUS HEAD(AFT)

SUB-COSTS

    JOINING HEAD(AFT
        TO CASE WITH INTEGRAL HEAD   83.83

TOTAL   83.83

FIGURE 9. COMPUTER OUTPUT DATA FOR ALPHA PRESSURE BOTTLE

TOTAL COST OF METAL ROCKET MOTOR

	ALTERNATIVE I (CASE AND HEAD(FORE) SEPARATE)		ALTERNATIVE II (CASE AND HEAD(FORE) INTEGRAL)	
FORMING AND MACHINING				
CASE	ROLL BENDING	580.89	DEEP DRAWING	228.02
HEAD(FORE)	DEEP DRAWING	183.14		
HEAD(AFT)	DEEP DRAWING	71.04	DEEP DRAWING	71.04
JOINING		143.49		83.83
HEAT TREATING		169.60		169.60
MOVING		51.01		51.01
HYDRO TESTING		27.92		27.92
TOTAL		1227.09		631.42

FIGURE 10. COMPUTER OUTPUT DATA FOR ALPHA PRESSURE BOTTLE

Reprinted with permission from the Society of Plastic Engineers.

COMPUTER ESTIMATION OF MOLDING AND TOOLING COSTS FOR WOODGRAIN PLASTIC PARTS

John E. Johnson

CONSUMER PRODUCTS DIVISION

The Magnavox Company

Fort Wayne, Indiana

INTRODUCTION

The estimation of plastic molding and tooling costs are important steps in the development of many plastic parts. For example, molding costs and material costs determine piece-part cost which often is a deciding factor in determining if a part can be made profitably or not. Piece-part costs are also important in deciding which of several alternative designs should be used. Tooling cost estimates are needed to prepare future tooling budgets and check on vendor tooling quotes. The combination of molding costs and tooling costs determine the number of cavities that result in the lowest cost for a particular part demand.

Molding cost is a function of machine cost rate, cycle time, and the number of cavities in the mold. This function can be represented by the following equation:

$$MC = \left(\frac{MCR\$}{HR}\right)\left(\frac{CT\ SEC}{CYCLE}\right)\left(\frac{HR}{3600 SEC.}\right)\left(\frac{1}{N}\right)\left(\frac{CYCLE}{PARTS}\right)$$

MC = Molding cost per part
MCR = Machine cost rate
CT = Cycle time
N = Number of mold cavities

The determination of the machine cost rate is not a difficult problem since it depends strictly on the machine size used to mold the particular part. The machine size is determined by the size of the part and the number of mold cavities. The cycle time is a much more difficult factor to estimate because it is a function of many variables such as plastic type, wall thickness, part tolerance and part size. At the present time, there is no exact way of calculating the cycle time. In spite of this, an experienced plastic cost estimator will be able to estimate costs to within 10% of the actual costs in most cases.

Tooling costs are estimated in two different ways. If very accurate estimates for tool authorizations or quotations are needed, a detailed analysis of machining and material costs must be made. This is a time consuming process that requires a person with a considerable amount of tool and die experience. He must be able to picture the complete tool in his mind and be able to estimate the man hours required to machine the various parts of the tool. There are few people qualified to make estimates of this type.

Tool estimates that are not as accurate as the above are often needed for feasibility studies and budget estimates. These estimates are made by an entirely different method than that described above. For example, a new tool to be estimated is found to be similar to a tool made last year. The new tool will be assumed to cost the same as the old tool plus a factor added for inflation of tool costs.

From the above discussion, it is apparent that molding and tooling costs are often made by simply considering costs of similar parts made in the past. The problem is that usually no old part can be found that is exactly like the new part. Because of this it is necessary to interpolate between levels of parameters that determine part costs. For example, if it is necessary to estimate the cost of a plastic overlay that is 12 in. x 18 in. x .5 in. and the nearest previously molded part was 12 in. x 12 in. x .5 in. it is necessary to correct for the smaller size. It is possible that it might be necessary to interpolate with respect to several parameters such as thickness, length, plastic types, etc. In a case with more than two variables, it becomes impossible to make corrections because of the possibility of having interactions among the variables.

One solution to the problem of estimating molding costs and tooling costs is to use a computer. The computer is able to estimate these costs by generating regression equations from data on parts molded in the past. From these regression equations and data on the new parts to be estimated, the molding and tooling costs can be estimated.

A simple example can best illustrate how these regression equations can be generated.[1] Suppose it was necessary to develop a regression equation that would predict molding cycle time. One important variable that determines molding cycle time is part weight. If a plot of cycle time versus part weight were made, it might look like Figure 1. The simplest type of regression equation that could be used to fit this data would be a straight line. This type of equation would predict about 69% of the variance and result in an average prediction error of 16%. If a quadratic equation is used, the average error could be reduced to 14.6%. There is still, obviously, other variables that are causing the data to deviate from the regression equation. If another variable such as thickness is added, even more of the variance can be explained. Figure 2 shows a two variable model with no interaction between thickness and part weight. The difference in cycle time between parts with a thickness of .150 and .125 is constant regardless of the part weight. A more realistic model is shown in Figure 3. An interaction term is added to take into account the simultaneous effect of thickness and weight on cycle time. If enough data is available even more variables can be added to further reduce the errors and increase the percent of explained variance.

Computer programs are available for developing regression equations in a manner similar to that described above.[2] The computer program will determine, from a list of several possible variables, which will explain the most variance. This variable is introduced into the model and the remainder of the variables tested to see which is the most important variable. This process is continued until the addition of more variables does not result in significant reduction in explained variance.

The first attempts at developing a computer program to estimate molding and tooling costs dealt with several types of plastic parts. About one hundred parts consisting of twenty different types of parts were first analyzed. It was found that it was necessary to specify thirty or forty variables to adequately describe the parts. In order to analyze this many parts, it would have been necessary to have data on several thousand parts. To reduce the number of parameters, the data was split into groups according to the type of part. For example, molding and tooling cost regression equations were developed for each type of part such as television fronts, doors, legs, overlays, etc. The number of parameters that could be used in the regression equation was a function of the quantity of data on past parts that were available. For parts that had only ten data points, it was possible to use only two or three variables to form the regression equation. For parts that had forty data points, it was possible to use as many as six or seven variables.

This paper will be concerned with the estimation of molding and tooling costs for woodgrain parts since more data was available on these types of parts than any other. Figure 4 shows a typical woodgrain overlay of the type considered in this paper.

MOLDING COST ANALYSIS

The first problem was to calculate the machine size needed to mold a given woodgrain part. A schematic of the program used to calculate machine size is shown in Figure 5. The logic used by this subprogram closely matches the thought process used by a tool analyst in determining the necessary machine size. The selection of the minimum sized machine to mold a plastic part is based on three factors: part size, part weight and part area.

The machine size is first calculated on the basis of part size. According to data collected in our molding plant, the maximum part length was three inches less than the horizontal platen size dimension. The maximum vertical dimension was three inches less than the vertical tie rod spacing. The part size was compared with the tie rod spacing and the platen sizes of all the machines in our plant and the smallest one that could mold the part found.

The machine size was next calculated on the basis of part weight. Ninety percent of the maximum machine shot size was considered to be the maximum part weight that could be molded.

A clamping pressure of 1.4 tons/in.^{2} was used to calculate the minimum sized machine based on area. This low value for clamping pressure was likely due to the low injection pressure needed to push the plastic through the typical .150 inch wall sections. The largest of the above three machine size values was the one that would be needed to mold the part. From the machine size the machine hourly cost rate was determined according to the latest available plant data.

The cycle time necessary to mold a part is a function of the following factors:

1. Injection Time = f(part type,side,number of cavities).
2. Cooling Time = f(part thickness,part type).
3. Part Removal Time = f(part size,number of cavities).
4. Machine Open and Closed Time = f(machine size).

The experienced molding cost analyst would calculate or guess the time it would take to inject plastic according to the length of flow, thickness and type of plastic. In order to determine the length of flow, it is necessary to know the location of the gating points. Since there is no simple way of determining the gating system that should be used for a given part, it was assumed that the gating system was fixed by the type, size of part and number of cavities. In other words, if the computer is given data on the type of part, the size of part and the number of cavities, it is not necessary to directly find where the gate points will be located. Since high impact polystyrene was used for all woodgrain parts, plastic type was not considered a variable.

Cooling time is a function of the part thickness and of the part quality required. If the part must be straight and free from sink marks, the part will require a longer cooling time than if the part appearance was not important. For woodgrain parts, the straightness requirements are nearly the same for all parts of equal size and type.

Part removal time is a function of the size of the part. Large parts usually take longer to remove from the mold than small ones. Machine open and closed time is a function of machine size, which is in turn fixed by the size of the part.

Figure 6 shows a schematic of the above logic. A stepwise regression program was used to determine the exact relationship between the input variables and cycle time. The stepwise regression provided a means for developing a mathematical model that would predict the cycle time based on the above input variables. From the original input variables ten more were generated to account for the intractions among the variables. The variables used to develop the mathematical model were as follows:

1. M=part weight
2. N=number of cavities
3. L=part length
4. W=part width
5. D=part depth
6. M^2
7. N^2 quadratic terms
8. L^2
9. D^2
10. MN
11. NL
12. NW
13. ND — two way interaction terms
14. LW
15. LD
16. WD
17. LWD
18. NLD — three way interaction terms
19. NLW

A set of forty-three computer cards were punched with cycle time data collected in our molding plant. A stepwise regression computer program determined which of these variables explained the most variance and introduced it into the model. This step was repeated until the addition of more variables resulted in no significant reduction in the percent of explained variance.

TOOLING COST ESTIMATION

The tooling cost computer program estimates both the core and cavity tooling costs. The assumption was made that the cavity was produced in house and the core produced by an outside vendor. The cost of making the cavity is a function of the following costs:

1. Ceramic cost
2. Metal cost
3. Labor cost
4. Cleaning cost
5. Straightening cost
6. Heat trest cost
7. Back-up plate cost

The ceramic and metal costs were based on the enclosed volume of the cavity. The other costs were determined from data supplied by our casting laboratory. The above costs do not include the cost of the wood model or the cost of the silicone rubber mold used to make the ceramic.

To calculate the cost of machining the core, it is necessary to calculate the cost of steel required and the number of hours required to machine the holes, bosses, ribs, etc. Since it was not possible to find this type of information, some assumptions were made. The cost of machining the ribs, holes, etc. was assumed to be fixed by the size of part and the type of part. For example, the costs for machining a core for a woodgrain overlay was assumed to be fixed by the length, width and depth of the part. The cost of the steel is a function of the parts overall size. The amount of steel removed is a function also of the size and type of part. The type of ejector system also relates to the tooling costs. It was assumed that parts smaller than 450 square inches would be air ejector tools.

The stepwise regression computer program was again used to find a mathematical model that related part length, width, depth and part type to tooling cost. Because of the lack of input data for some of the part types, part enclosed volume and part type were the only variables used for some types of parts.

The following information is inputed to an IBM system 360 model 44 computer by means of a 2250 graphic input terminal:

1. Type of analysis
2. Type of part
3. Overall part length
4. Overall part width
5. Overall part depth
6. Overall part weight
7. Part projected area
8. Part demand
9. Date of estimate

ACCURACY OF ESTIMATES

To estimate the ability of the computer program to estimate cycle time and tooling costs, several parts were chosen at random and the prediction errors analyzed. The average cycle time error was 8%. A detailed analysis of the cycle time errors is shown in Figure 7. From the graph, it is apparent that there is a 20% chance of having an error greater than 10% and a 12% chance of having an error greater than 15%. Since the error in predicting machine cost rates is low, the same curve would approximate errors in predicting piece-part cost. Figure 8 shows the tool cost estimating errors for overlays. The average tooling cost error was 10%. There is a 40% chance of having an error greater than 10% and a 20% chance of having an error greater than 15%.

CONCLUSIONS

The computer programs described in this paper are now being used to make feasibility studies and budget estimates for woodgrain plastic parts. It allows engineers and scheduling personnel, inexperienced in estimating piece-part and tooling costs, to quickly obtain cost estimates based on simple part data obtained from blueprints. In order to obtain reasonable accuracy with this type of program, it is necessary that the parts of each type be geometrically similar to parts made in the past. The parts also must not be significantly larger or smaller than similar parts made in the past.

ACKNOWLEDGEMENTS

The author wishes to thank Mr. Dale Thomas and others in The Magnavox Company Tooling Section for their sincere help with this project.

BIBLIOGRAPHY

1. "Introduction to Linear Models and the Design and Analysis of Experiments," W. Mendenhall, Wadsworth Publishing Co. Inc., Belmont, California, 1968, pages 51-57.

2. "System/360 Scientific Subroutine Package-Version III - Programer's Manual," IBM Corp. 1970, pages 41-43.

3. "Statistics in Research," B. Ostile, Iowa State University Press, Ames, Iowa, 1963, page 33.

4. "Computer Adds Process Selection and Costing to its Repertoire," Plastics World, Feb.1971, 29-2, pages 38-42.

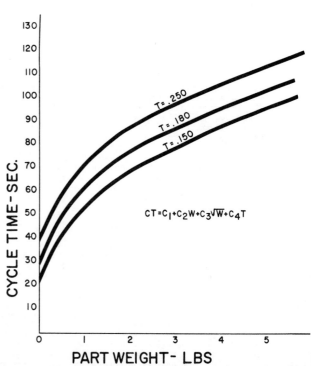

FIGURE 1 - LINEAR AND QUADRATIC REGRESSION EQUATIONS

FIGURE 2 - TWO VARIBLE MODEL WITH NO INTERACTION

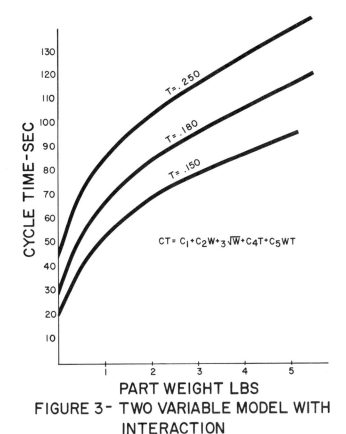

FIGURE 3 - TWO VARIABLE MODEL WITH INTERACTION

$$CT = C_1 + C_2 W + 3\sqrt{W} + C_4 T + C_5 WT$$

FIGURE 4 - TYPICAL PLASTIC WOODGRAIN OVERLAY

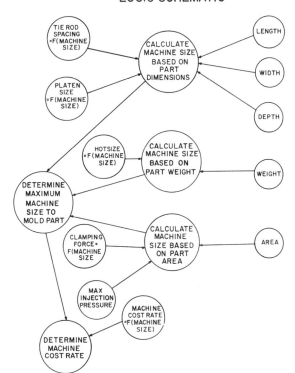

FIGURE 5 MACHINE SIZE ESTIMATION LOGIC SCHEMATIC

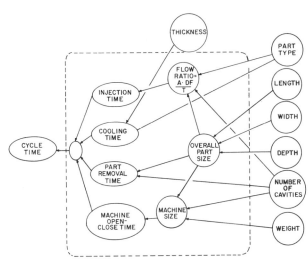

FIGURE 6 CYCLE TIME ESTIMATION LOGIC SCHEMATIC

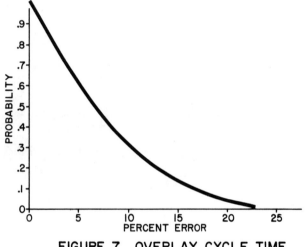

FIGURE 7 OVERLAY CYCLE TIME
CUMULATIVE PROBABILITIES OF EXCEEDING A GIVEN ERROR

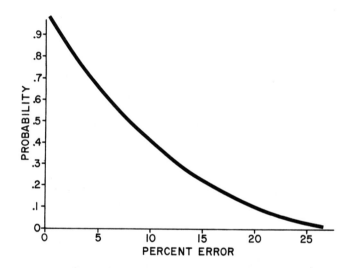

FIGURE 8 OVERLAY TOOLING COST
CUMULATIVE PROBABILITIES OF EXCEEDING A GIVEN ERROR

CHAPTER 3

PRO FORMA

Proforma In Manufacturing Cost Estimating

By Phillip F. Ostwald
Professor, Mechanical and Industrial Engineering
University of Colorado

ABSTRACT

While cost estimates are useful for many purposes, they may be unable to answer management questions directly, or may provide poor quality answers. Four methods of product estimating are graded on 22 inner comparison factors. Using this paper as a guide, a company will be able to give thoughtful study to improvement of their own method of product estimating.

INTRODUCTION

The output of a cost estimator is a cost estimate. This information is required for many purposes, chief of which is product pricing. That the estimate be factual and timely is recognized, but unfortunately the estimate may be unable to answer other management questions and additional work or re-estimating is required. Or the work may be estimated, analyzed, and arranged in a way to give incorrect answers. A proforma (a latin word meaning "for the sake of form") is the company's recognized and official cost estimating form that summarizes the total direct labor, material, and overheads of various kinds. The proforma in manufacturing cost estimating is also called the "recap sheet" or "summary" and is the single most important document of the company's cost estimating efforts. It is most often approved by the owner, or president or another senior executive officer. A great deal of attention is given to the cost values as provided on summary cost estimating form, but little thought is directed to advantages and disadvantages of a method of cost estimating.

Our paper reports on this comparison. We provide no final recommended proforma for cost estimating as the choice is a compromise of many requirements. Initially, we describe the reasons for variety in cost estimating. Continuing, inner comparison factors are collectively discussed. The methods, and four are provided, are discussed with examples. Twenty-two inner comparison factors are graded. Finally, tabular definitions for the factors are listed.

BACKGROUND TO PROBLEM

With over 100,000 firms that manufacture a product in the United States, it must be recognized that there are many ways to estimate a product. Indeed, surveys by the author have shown that some companies may have several approaches. Given that no one best way exists to estimate a product, there are, however, four approaches that characterize this variety. We identify these methods as (1) Operation, (2) Department, (3)

Variable Cost, and (4) Learning Curve. Which general method to adopt can create confusing choices. Advantages gained by one method are denied by another. For instance, a mass-produced product that is similar over the years offers advantages in analysis, accuracy, volume-sensitivity that would not be found with procedures used by job-shop organizations. Or opportunities offered by the procedures to estimate spare parts using the original product estimating scheme may be awkward to provide accurate and readily determined values for cost reduction. It is apparent that the proforma must satisfy many objectives, although not with the same success. The purpose of this paper is to examine many factors and grade their success within the four proformas for product estimating. A firm will develop its cost estimating procedures to satisfy many requirements and will compromise the purposes that the cost estimates can offer.

INNER COMPARISON FACTORS

Twenty-two factors are examined and graded on an A, B, C, and D basis. Each factor is defined and collectively listed in Table 1. The grading isolates that factor from the other factors even though there may be an association. The A, B, C and D grade indicates a judgment and is no more precise than a relative index.

The factors are gathered into four categories, or

1. Organizational Purpose

2. Estimating Costs and Preparation

3. Marginal Propensity

4. Versatility

Organizational purpose deals with the primary function of an estimate as it relates to the type of an organization. A governmental procurement method and a consumer product method are dissimilar and have differing objectives. The "cost of estimating cost" is often a major expense. If preparation were simplified, we presume that the cost of estimating cost would not receive the criticism that it often does.

Marginal propensity deals with the robustness of the method to reflect marginal cost concepts. Marginal cost is, of course, the difference in total cost between neighboring units of production, say the 600 and 601st. Most methods do not show this neighboring difference, but there is an inclination or propensity. Many company methods use "average" cost values, which is a poor concept when compared to the "marginal-propensity" concept.

The primary purpose of a cost estimate is to provide the full cost for manufacturing, development, and sales. Secondary purposes are many, but the method may not disclose the desired secondary information and simple or extensive additional work may be required. Five inner comparison factors are reviewed with respect to versatility of the method of cost estimating.

Table 2 summarizes the grading of the four factors. The grade A, B, C, or D are relative to each other where A is superior and D means below average for the inner comparison factor.

OPERATION METHOD

The operations process sheet is the originating document. Once the op sheet is underway or finished, cost estimating can proceed. Figure 1 is an example of the requirements for information and shows the flow of information from the components to subassemblies to final assembly. The measured and policy information requirements are setup hours, hours/100 units, machine labor rate in $/hour, and machine labor and overhead rate. The product of these quantities provides a labor and overhead cost for the operation. This calculation performed for each operation is the reason for the name of the method. The calculations are "per unit" rather than "per lot" and are carried along starting with the component column total for material and labor and overhead. Note that $0.15 and $2.205 (material and labor and overhead costs) from part number 672 are posted against part 642 which is a subassembly. If material and labor and overhead are contributed to the process, these costs would be calculated for each subassembly operation. These costs are eventually posted to the final assembly where G & A costs, engineering costs, and selling costs are considered. Once the subassembly is finished, the costs are added vertically downward and the cost includes material and labor and material cost for the components and the subassembly.

For instance, operation 10 of part number 642 uses these calculations, or

$$\text{Unit estimate} = \frac{\text{Setup hour}}{\text{Quantity}} + \text{Hour/unit}$$

$$= \frac{0.1}{100} + .001 = 0.002 \text{ Hour/unit}$$

$$\text{Labor \& Overhead unit/cost} = (\text{unit estimate}) \times (\text{labor rate})$$
$$\times (\text{labor+overhead rate})$$
$$= (0.002)(4.35)(1.70) = \$0.015$$

Note Figure 1 for a description of this method.

DEPARTMENT COST METHOD

While route sheets are useful for the Department Method, they are not as necessary as for the Operation Method. If op-sheets are available, the hours per lot are summed across the op-sheets for each department and their total entered on the proforma. If op-sheets are unavailable, other historical information or estimates are substituted. The collection of these lot hours versus various departments is the reason for the title of the method. Extension by the average wage rate for the department gives a department labor cost. Other features like a deduction for learning curve can be managed. Other calculations are required, specifically for plant overhead, G&A costs, materials and materials overhead, custom engineering and overhead, and selling. An example of this method is provided by Figure 2.

VARIABLE COST METHOD

This method is titled "Variable Cost" because of the emphasis on finding full variable cost. Measured values as found by the op-sheet would be mandatory to indicate the direct labor and direct material costs. But once these quantities have been calculated other analyses are necessary. Information such as variable manufacturing overhead, freight, irregular merchandise allowance, etc., are required to complete the Variable Manufacturing Costs section of the proforma. Additionally, variable marketing and administrative costs are collected and these kinds of information may not be generally available to manufacturing. The concentration of effort to identify the variable costs is the reason for the name of the proforma.

Fixed overhead are separated into standby and product. This distinction allows for fixed costs that are necessary for basic operation even if no product is made, and fixed cost identified with a specific product. The contribution method of pricing is required. See Figure 3 for an example of this method.

LEARNING CURVE METHOD

A major feature of this method is the incorporation and dependence upon the learning curve. While Figure 4 deals with fixed and variable cost categories for the first unit, any unit can be identified as the base unit. The learning curve calculation can handle a variety of contexts, such as unit linear line or cumulative average linear line assumptions and reference [1] can be studied for application. The primary estimate divides selected costs into fixed and variable. The selection of a learning slope depends upon the type of cost. For instance, assembly will have more learning than a punch press department. Learning curve tables, computer or calculator programs, or formulas are commensurable approaches to extending costs.

---
1 Cost Estimating for Engineering and Management, Phillip F. Ostwald, Prentice-Hall, Inc., Englewood Cliffs, N.J., 1974.

CONCLUSIONS

The cost estimate is the center of information useful for many decisions. The ability of cost estimators to deal with management questions depends upon the cost estimate. This opportunity and flexibility furthermore depends upon which method of product estimating is used. An inner comparison is made of 22 factors for four major product estimating methods. Using this paper as a guide, a company will be able to give a thoughtful study to the improvement of their own method of product estimating.

Table I. Definitions of Inner Comparison Factors

<u>Accuracy</u> = $(C_e/C_a - 1) \times 100$ where accuracy is also called percentage error, C_e = estimated cost/unit, and C_a = actual cost/unit. A grade of A is considered to be more accurate insofar as the method is concerned.

<u>Audit path</u> deals with the capability to trace costs using a proforma system beginning with elementary cost entries to the finding of the final full cost. A grade of A implies the better method for tracing of costs.

<u>Capture percentage</u> = (number of estimates won) ÷ (number of estimates made) X 100. A job shop would have a lower value, for example, than a manufacturer of a proprietary product. A level of 25% was adopted arbitrarily. A grade of A suggests the better method to use if the capture rate is less than 25%.

<u>Companion estimates</u> are special-purpose proforma, i.e., material estimate, tool estimate, or engineering estimate. They are included in the final proforma product cost estimate. A grade of A indicates that the method requires special companion estimates in the method.

<u>Consumer product</u> is one purchased by a consumer, i.e., TV's, shirts, foodstuffs, or toothbrushes. A grade of A indicates that the method is relatively more suitable for consumer products.

<u>Cost reduction</u> is the ability to reduce manufacturing cost using a method. A grade of A suggests that the method provides this information directly in a relative and easy way.

<u>Desperation estimating</u> is the act of finding "bare bones" cost for the purpose of maintaining plant operations. Hardship conditions may create a temporary policy of desperation estimating. A grade of A indicates

relative ability to reduce full costs and quote a minimum value to obtain business.

Ease of preparation deals with the requirements and costs for estimating the product. A grade of A would be less costly than a D.

Federal procurement regulation suggests the relative compatability of the method to Federal practices and laws. While all methods may be satisfactory, Federal contractors are encouraged to follow special reporting forms, i.e., Form 633.

Full variable cost = price - profit - fixed cost.

Industrial products sell to industry and examples are turbines, machine tools, and conveyors. A grade of A indicates the compatability of the method to industrial products.

Intensity, whether labor, material, capital, development, or a combination of these, is the ratio of the cost to full cost, i.e., labor cost/full cost X 100; material cost/full cost X 100, etc. A grade of A indicates that the method is more calculation of these ratios.

Job shops prepare an estimated cost and quote a price for a customer request. If the estimate wins, the company orders material, tooling, and spends money to fulfill the requirements. If the company fails to win, no further effort is expended. Other terms are "contract" or "custom shop". A grade of A suggests that the method is generally more suitable for job shops.

Joint costs exist whenever from a single source of material or process there are produced units of goods having various unit values. A grade of A indicates a relative capability by one method which would be superior to others.

Marginal propensity is the tendency to provide the added costs of producing one additional unit of product without additional fixed cost investment. We presume that the proforma has the capacity to indicate the marginal cost. A grade of A identifies a method more likely to accurately provide marginal cost information while a D grade would provide a poor substitute.

Make vs. buy is the comparison of in-plant to purchase costs under full or partial shop loads. A grade of A implies relative simplicity.

Proprietary product is a self-designed or licensed product that a company will manufacture. A grade of A identifies a method suitable to a proprietary product.

Re-estimate deals with the difficulty of updating an estimate for changes in design or cost. A grade of A suggests that the method is relatively compatible or simple to update.

Setup identification indicates that setup is obvious rather than hidden or lumped with other information. A grade of A implies that the setup values are visible, i.e., specifically stated.

Spare parts reveals the estimating efforts using the proforma to identify costs for spare parts. A grade of A would indicate that the proforma is compatible to the finding of spare parts costs.

Volume sensitivity points up the relative simplicity of calculations to find volume effects. A grade of A would indicate that the method was responsive to volume variation.

INNER COMPARISON FACTOR	OPERATION METHOD	DEPARTMENT METHOD	VARIABLE COST METHOD	LEARNING CURVE METHOD
I. Organizational Purpose				
Consumer product	C	C	A	D
Federal procurement regulation	C	B	D	A
Job Shop	B	A	D	D
Industrial product	B	A	D	C
Proprietary product	A	A	A	C
Re-estimate	D	A	D	C
II. Estimating Costs and Preparation				
Capture percentage 25%	C	A	C	C
Comparison estimates	C	A	D	B
Ease of preparation	B	A	C	C
Ease of understanding	B	A	D	C
III. Marginal Propensity				
Accuracy	A	C	B	C
Desperation estimating	C	A	A	B
Full variable cost	D	A	A	C
Intensity	C	A	D	B
Marginal propensity	C	A	A	B
IV. Versatility				
Audit path	A	C	D	C
Cost reduction	A	C	C	C
Joint cost	B	C	C	C
Make vs. buy	A	C	D	D
Setup identification	A	C	D	C
Spare parts	A	C	D	C
Volume sensitivity	C	B	A	A

Table 2. Comparison of factors for four methods of product estimating

Form — Part number 672

Part number 672
Quantity 100
Estimator Led
Date 3/21/-
Estimate expires on 6/19/-

Work center	Oper. no.	Description (List tools...
Stock		Raw materi...
7/00	10	Shear
8/300	20	Tape Pierce N/C Tape
47/03	30	Csk 18-0... 4-0...
853	40	Degrease +
71	50	Brake (2
71	60	Brake (2
47/03	70	Ream 24-0.125 +000 -000 + 4-0.189 holes
853.1	80	Degrease + deburr
956	90	Heliarc 4 corners
234-2	100	Grind welds

Setup hrs.	Hrs/100 units	Unit estimate	Labor rate	Overhead rate	Labor & overhead
0.3	2.700	0.030	2.90	2.15	0.187
0.2	0.600	0.008	1.95	1.90	0.029
0.2	3.500	0.037	3.06	2.35	0.265
0.1	2.000	0.021	1.95	2.35	0.096
Component total				$0.150	$2.205

Form — Part number 642

Part number 642
Quantity 100
Estimator Led
Date 3/22/-
Estimate expires on 6/19/-

Part name Chassis, Subassembly
General notes: Subassembly 2 Components

Component part number	Material	Labor & overhead
672	$0.150	$2.205
473	0.190	2.701
Subtotal	$0.340	$4.906

Work center	Oper. no.	Description of operation (List tools and gauges)	Setup hrs.	Hrs/100 units	Unit estimate	Labor rate	Over-head rate	Material	Labor & overhead
921-6	10	Deoxidize	0.1	0.100	0.002	4.35	1.70		0.015
958	20	Spotweld 10 spots Part 672 to 473	0.4	3.000	0.034	6.10	1.25		0.467
47/03	30	Drill thru (1) hole B Csk (1) hole B	0.3	2.300	0.026	5.75	2.5		0.321
911	40	Degrease + deburr side	0.2	0.600	0.008	4.45	1.90		0.068
913	50	Chromate	0.1	0.300	0.004	4.40	2.00		0.035
500-1	60	Silk screen	0.3	0.400	0.007	5.10	2.10		0.075
Total								$0.340	$5.887

Move to inventory —

Form — Subassembly

Subassembly part number	Material	Labor & overhead
642	$0.340	$5.887
731	$4.210	$1.071
872	$6.020	$15.211
127	$4.387	$4.380
Subtotal	$14.957	$24.549

Labor rate	Over-head rate	Material	Labor & overhead
5.27	2.17	0.070	0.435
4.95	3.27	0.170	0.599

1. Cost of goods manufactured: Total $15.197 $27.583
2. General & administrative costs @ 90% of 1 $38.502
3. Engineering costs prorated $33.825
4. Contingencies —
5. Selling Costs @ 40% of 1 $17.112

Total unit cost of manufacturing, development & sales $132.215

Figure 1

Manufacturing estimate for _____ products

Description: Speed decreaser – LH

Customer _____ Quote no. _____ Date _____
Base on quantity of 480 units During period of _____

Dept.	Description	Hr/Lot	Rate	Labor
103	Finishing	33.60	$5.60	$188.16
105	Machine shop	143.20	$6.00	$859.20
106	Miscellaneous machines	118.20	$5.75	$679.65
108	Precision assembly	28.80	$7.00	$201.60
131	Electronic assembly	68.05	$7.10	$483.16
125	Cable assembly	—		
233	Inspection	28.80	$7.20	$207.36
241	Stock room	48.00	$5.10	$244.80
	Subtotal	468.65		$2863.93
	Learning curve factor –8%			$299.11
	Subtotal – manufacturing labor			$2634.82
	Fringe labor cost + 28%			$737.75
	Total manufacturing labor			$3372.57

Item	Description	Unit	Lot
1.	Manufacturing labor (per above)	$7.026	$3372.57
2.	Manufacturing overhead @ 75% of item 1	$5.270	$2529.43
3.		$12.296	$5902.00
4.	General & Administrative @ 20% of item 3	$2.459	$1180.40
5.	Parts + materials per estimate	$2.321	$1113.80
6.	Overhead on parts and materials @ 15%	$0.348	$167.08
7.	Custom engineering labor per estimate	$8.330	$3998.50
8.	Custom engineering overhead @ 40%	$3.332	$1599.40
9.	Research and development per estimate	—	—
10.	Contingencies per estimate	—	—
11.		$29.086	$13,961.27
12.	Selling @ 25% of item 11	$7.272	$3,490.32
13.	Total cost	$36.358	$17,451.59
14.	Profit @ 13%	$4.727	$2,268.63
15.	Selling Price	$41.085	$19,720.22
16.			
17.			
18.			
19.			
20.			
21.			
22.			
23.			

Cost approvals		Price approvals	
Estimated by	_____	Accounting	_____
Approved by	_____	Product manager	_____
Division head	_____1____	Dir. of service engrg.	_____
	_____	General manager	_____

Figure 2

PRODUCT COST ESTIMATE SUMMARY

Preliminary or detail _____
Product line _____ Retail price $0.50 Retail mark-up 44.5%

ITEM	AMOUNT	PER CENT
LIST PRICE	$0.2775	100.00
VARIABLE MANUFACTURING COSTS:		
Direct material	0.741	26.70
Direct labor	0.0243	8.76
Variable manufacturing overhead	0.0550	19.82
Freight to warehouses	0.0042	1.51
Irregular merchandise allowance	0.0022	0.80
Freight to customer	0.0021	0.77
TOTAL VARIABLE MANUFACTURING COSTS	0.1619	58.36
VARIABLE MARKETING COSTS:		
Trade discount allowed	0.0028	1.01
Field allowances	0.0001	0.04
Promoting allowances	0.0035	1.26
Royalties	—	—
Variable selling expenses	0.0031	1.11
TOTAL VARIABLE MARKETING COSTS	0.0095	3.42
VARIABLE ADMINISTRATIVE COSTS:		
Cash discounts allowed	0.0050	1.80
Variable administrative expenses	0.0040	1.44
TOTAL VARIABLE ADMINISTRATIVE COSTS	0.0090	3.24
TOTAL VARIABLE COSTS	0.1804	65.00
STANDARD PROFIT CONTRIBUTION	0.0971	35.00
STANDBY FIXED COSTS:		
Manufacturing	0.0208	7.50
Selling and marketing	0.0077	2.77
Advertising	0.0002	0.07
Administrative	0.0069	2.48
TOTAL STANDBY	0.0356	12.82
PRODUCT FIXED COSTS:		
Manufacturing	0.0066	2.38
Selling & marketing	0.0039	1.40
Advertising	0.0066	2.38
Administrative	0.0200	7.24
TOTAL PRODUCT	0.0371	13.37
STANDARD EARNINGS	0.0244	8.79

Figure 3

TABLE 9.4

LEARNING CURVE APPROACH TO COST ESTIMATING

Selected Costs	First Unit Cost Estimate		Estimated Learning Slope	150-Unit Cumulative Factor	150-Unit Cumulative Cost	Total Line Cost
	Fixed	Variable				
Manufacturing	—	$4700	75%	30.93	$145,371	$145,371
Raw material	—	900	90	82.15	73,935	73,935
Engineering	$43,000	380	85	59.89	22,758	65,758
Tooling	5,100	240	70	21.97	5,272	10,372
Quality control	1,600	230	75	30.93	7,113	8,713
Equipment	8,400	330	90	82.15	27,109	35,509
Other direct costs	2,500	45	80	43.23	1,945	4,445
			1. Total selected costs			344,103
			2. Other indirect costs			28,000
					Subtotal	370,103
			3. Fixed and miscellaneous charges @ 0.28			103,628
			4. Distribution and administrative costs @ 0.21			77,722
			5. Contingencies @ 0.08			29,608
			6. Selling costs @ 0.02			7,402
					Subtotal	588,463
			7. Profit @ 0.18			105,923
					Total price	$694,386

Figure 4

Figure 4 is from <u>Cost Estimating For Engineering And Management</u> (page 281). Copyright 1974 Prentice-Hall, Inc.

Estimating Labor and Burden Rates

J. G. MARGETS
IBM Corporation

REDUCED TO ITS BASIC ELEMENTS, the price paid for each procurement consists of the seller's projected performance (unit hours) and his selling price (labor and burden rate) for each unit hour. Content and significance of unit hours are generally understood and negotiable for each procurement. To effectively negotiate the major portion of a pricing proposal, however, the buyer must fully understand the content of labor and burden rates, and the potential impact of each element on cost.

Because of the nature of its content, estimating a labor and burden rate requires application of several values derived from the experience and best judgment of the estimator. Information in this article is offered as a discussion of an outline of a labor and burden rate structure, shown in the accompanying box. Some of these considerations, if applied, should influence and support judgment values.

An accurate definition of the procurement is essential to a good estimate. The estimate should cover the same total quantity of goods/services and time span as the procurement. Also, the goods/services to be purchased and the time span of the purchase or delivery must be quantified to project costs and properly allocate them in the estimate.

Direct Labor Costs. This item should include the wage-only cost of all direct labor required to produce the total procurement requirements. Direct labor is defined as hands-on labor performing an operation relative to the production or assembly of the product. Most procurements involve products requiring multiple operations. Each operation contributes to the total cost in proportion to its wage rate and required total man-hours. This contribution also includes allowances for yields and possible learning curve applications.

Wage rates for various skills may be drawn from recent experience, U.S. Bureau of Labor statistics, state employment offices, or area newspapers. Final selection of a wage rate could be influenced by a knowledge of available resources relative to union or nonunion labor, temporary or full time labor, labor turnover history, and the effect of such factors on the normal average wage rate.

Direct Material Costs. After factoring the procurement requirements for yield and scrap, the material costs should be calculated. These should include the costs that the seller will incur for transportation and purchased services, including plating, treatment, processing, etc. Allowances for purchasing, handling and warehousing are inputs for other sections of the estimate.

Unusual price and stock situations can occur where material costs (precious metals for example) are a major portion of the total procurement. In such cases, it is advantageous for the buyer to isolate associated material costs, estimating their values and markups separately.

Indirect Labor Costs. Manufacturing overhead is a summary of various manufacturing related costs such as indirect labor, benefits and occupancy, resulting from the procurement in support of the direct labor that is required. Each entry should be itemized in detail for future reference. Indirect labor costs should include

---

OUTLINE OF A LABOR AND BURDEN RATE STRUCTURE

1. Direct Costs
   A. Labor
   B. Material

2. Manufacturing Overhead
   A. Indirect Labor
   B. Benefits
   C. Occupancy (rent, depreciation, utilities, taxes, insurance, maintenance, supplies and transportation)

3. General Works and Administration
   A. Salaries
   B. Services
   C. Supplies

4. Divisional Expense

5. Profit

6. Total Program Dollar Value

7. Labor and Burden Rate

---

the wage or salary cost of nonmanagement support personnel that can be identified with the program to the extent that they support this particular procurement.

Typical of the functions contributing indirect labor support are Manufacturing, Production Control, Quality Control, Maintenance and the various branches of Engineering. Wage and salary values that are applied will be influenced by a knowledge of industry average rates, and the relative cost of providing adequate support from available resources.

Benefits. A significant addition to wage and salary costs is the combined benefits—both mandatory (required by law or contract) and elective—incurred by the seller as an employer. Mandatory benefits would include items such as old-age, disability, survivors' and health insurance, and unemployment insurance. Pension plans, supplementary insurance, hospitalization plans and others might be mandated by labor contracts.

Elective benefits would include any of the above provided but not mandated by law or contract. Additional items would include vacations, holidays, lost-time compensations, bonus plans, tuition refunds, counseling services, cafeterias, awards and numerous others.

Values for government mandated benefits are generally based on specific percentages or wage-salary dollars, and should be available from the area personnel department or state labor office. Contract mandated and elective benefits can be estimated quite accurately if reasonable values are assigned to an itemized listing of applicable benefits.

Particular attention should be given to the different application of specific benefits to hourly versus salaried personnel. Typically these benefits would include holidays, vacations, and lost time compensation that could be considered additive to base hourly costs, but nonadditive to (included in) salaried costs. Social Security costs paid by the seller as an employer are applied to a fixed dollar portion of each employee's earnings. Virtually all of the direct labor costs will have the current percentage of tax applied, but the same percentage is applied to only 50 to 75 percent of the direct labor-salaried costs.

Occupancy. This is a summary of various manufacturing-related costs identified with and allocated to the direct costs associated with a procurement. Costs in this category are generally chargeable to each procurement on the basis of square footage of factory and administrative space utilized, or on the basis of the procurement's effect on total sales dollars.

It is necessary to determine the square

footage of factory space associated with the procurement for shipping-receiving, staging, manufacturing, materials storage, etc., and the administrative space normal to such an operation. Some of the areas will serve multiple procurements and should be proportionately allocated to individual procurements. One approach to estimating total square footage requirements involves the use of a familiar installation as a model. With the use of this method, any resulting variance from actual requirements will be less significant than by the application of rule-of-thumb values.

Rental costs for space, if applicable, should be analyzed for the exact coverage that is included. The rental costs may well include items that are covered elsewhere in your estimate, such as utilities, insurance and taxes. Only an understanding of the overall coverage can guide the estimator to avoid duplication. Building depreciation would apply in lieu of rental costs. If the specific data is lacking for a particular industry or supplier, an estimate could be based on the current replacement cost depreciated on a straight line basis over 20 years.

Costs in the tools-equipment depreciation category include all specific capital tooling required as a part of the procurement, and other depreciable assets required in the direct and support areas associated with the procurement. Estimates can use original costs or current replacement costs of each item, depreciated over an appropriate time period. Normally this depreciation is 10 to 15 years for standard machine tools, five years for office equipment, and one to seven years for special machine tools, depending upon their applicable procurement and technological life. Allocation of these costs should be in proportion to their support of a particular procurement versus other procurements.

Costs of electric, gas, water, steam and similar items should be included in estimates. The utilities included are those normal to the industry and geographical area, with particular emphasis on high consumption utilities peculiar to the procurement. The model-installation approach can aid in establishing a base for these costs.

All local, state and federal taxes should be estimated using a base established by the typical real estate, sales, and profits. Insurance costs should be estimated to cover the buildings, contents, and liability coverage.

An excellent source for basic tax and insurance cost data is the financial personnel within your company. Also, financial reports, assessor and tax offices, and insurance companies can provide such information. The tax and insurance costs should be allocated in your estimate on the same basis as the rent and depreciation costs.

Facilities maintenance (another occupancy item of manufacturing overhead) in-

---

Review procedures can eliminate all items of questionable value in the quote

---

cludes the costs of labor, supplies, and services that are normally required for the upkeep of buildings, offices, and distribution of utilities. Tools and equipment maintenance includes the costs of labor, supplies and services to maintain and repair the items of production equipment that are to be used relative to a procurement.

Facilities maintenance costs should be allocated in your estimate on the same basis as rent and depreciation costs. Tools and equipment maintenance costs are included to the extent that they support a particular procurement in comparison to other procurements.

Estimates should also include the expense of all tools, process related supplies (cleaning solutions, lubricants, plating chemicals, etc.) and factory supplies (aprons, gloves, timecards, etc.). Factory supplies can be estimated relative to the labor content of the procurement with reasonable accuracy as they generally represent a minor portion of the total procurement costs.

Allowances must be made in the estimate for transportation costs to be incurred by the seller relative to his procurement of tools, equipment and supplies. If the seller is to prepay the shipment of finished goods to you, the costs for transportation are recoverable and should be included in your estimate. Gross shipping weights must be calculated for each significant item. Shipping costs are usually available from the purchasing, traffic or accounting departments.

General Works and Administration. This category includes indirect expenses such as salaries, services and supplies that can be classified as the cost of being in business. Ideally, these costs should be only charged to a procurement to the extent that they support it. However, it is not uncommon for the seller to allocate these indirect expenses to each procurement in a proportion that is a percentage of his total sales.

Salary estimates should be based on the support required for a procurement relative to the officers, management, sales, legal, purchasing, finance, clerical, security and other applicable personnel. A detailed listing should be prepared identifying the type, quantity and dollar value of allowable support. Whenever the specific values cannot be identified to a procurement, estimates are often based on a proportionate share of the estimated total costs. Salary values are applied for each entry, and are influenced by a knowledge of the cost of available resources and overall industry averages.

Service costs include advertising, travel, communications, legal and similar items. Except for unique cases that require unusual expenditures, reasonable accuracy can be achieved by estimating the probable total costs and allocating a percentage of total costs to a procurement.

Supply costs include items normally associated with offices, sales, accounting and engineering. This category usually represents a nominal dollar value that can be similarly estimated and allocated as services.

If a procurement involves a commodity or industry whose technology requires continual research and development activity, estimates for this category should include allowances for salaries, services and supplies under the General Works and Administration heading. It is possible that a procurement may be placed with an industry engaged in research and development activities that are not related to your commodity. While the seller may expect your procurement to absorb some of his costs for his overall activity, your estimate should address only costs that can be directly identified with your requirements.

Divisional Expense. If the supplier is a division of some parent company or corporation, as an income-producing segment, the supplier must share expenses incurred by the parent organization. These expenses include salaries, services, supplies, and facilities related to officers, corporate staff,

Two profit pictures are important — yours and your vendor's

travel, advertising, benefits and other items. The supplier generally recovers such expenses by applying a percentage of 5 to 15 percent on his sales dollars.

Application of this percentage to an estimate for a procurement should be limited by the probability of sourcing a procurement with a supplier that is established as a division. Generally, it is difficult to identify corporate expenses if you do not have intimate knowledge of the corporation and an analysis of its financial reports (an approach that should be pursued if the potential exposure warrants the effort that is involved). It is possible to estimate reasonable costs using the General Works and Administration costs as a base and estimating the additions for similar Divisional Expense items.

The combination of General Works and Administration with Divisional Expense costs should not greatly exceed similar costs for a supplier that is not burdened by such supporting expenses. Unless this dollar value represents some contribution to a procurement, you should view it as an unnecessary addition—unnecessary to the extent that your requirements can be produced without incurring such costs.

Profit. The price paid for a procurement includes the return (profit) deemed reasonable by the supplier for his investment (projected costs). The importance of a reasonable profit cannot be overemphasized as it contributes to the success of the supplier, and enables him to continue his operation and expand his capabilities. Procurement estimates should include a reasonable additional percentage for this profit. This percentage should be applied to those items of cost which justify the addition for profit.

Financial reports can be helpful in determining the normal percentage of profit before taxes for a given industry. The minimum allowances, assuming realistic cost targets, should be the current cost of borrowing the total dollar requirement for the specified time span.

Normally, the profit percentage selected is applied to the subtotal of items 1 through 4 inclusive of the rate structure shown in the accompanying box. Any exceptions would include a portion of your estimate that, in your opinion, justifies a greater or lesser percentage. A greater percentage may be justified where a contingency for high risk is required. Lower percentage additions may be justified where rapid dollar turnover results in a compounding profit.

Total Program Dollar Value. This total of items 1 through 5 in the rate structure outline represents the estimated total dollar value of a procurement. It represents the dollar commitment that you must make for a specific quantity of goods or services over a prescribed time period.

The total program dollar value divided by the quantity of goods to be purchased will determine the estimated unit price or dollars per piece. Based on prior experience you will often have a basic understanding as to the probable unit price. If the estimate produces a significant variance, the estimate should be reviewed in detail for revision or confirmation of the input information. Reviewing the estimate with key engineering, finance and administration personnel within your company will isolate any questionable values and establish a greater level of confidence.

Labor and Burden Rate. The total program dollar value divided by the total direct labor man-hours will determine the estimated labor and burden rate. This value represents your estimate of a fair and reasonable charge-out rate for each estimated direct labor hour required to produce your requirements.

Quote Analysis. The value of performing a detailed estimate becomes apparent upon receiving quotes for a procurement from suppliers. The buyer cannot realistically assess the quotes that he receives without an estimate. Typical of the questions that he must consider are:

1. Is the lowest quote too high or low?
2. Does the lowest quote represent a fair value?
3. Will the lowest quote be as attractive as a higher, more negotiable quote?
4. To what extent can the quote be safely negotiated without jeopardizing the procurement and the supplier?

With the use of a detailed estimate, the buyer can determine his position relative to each of the quotes, and develop specific conclusions prior to the negotiations.

As an example to demonstrate the value of an estimate, assume that you have a procurement with requirements covering 100,000 pieces over a six-month period. Your estimated price is $2.25 each based on 0.2500 man-hours per piece at a $9.00 per hour labor and burden rate. An analysis of one quoted price of $2.70 each reveals that it represents a 20-percent or $45,000 variance on the total procurement.

An analysis of a summary of your estimate (accompanying TABLE) would negate the possibility of a $45,000 estimating error in any one-cost element. This would indicate that the supplier is seeking money that is not readily accountable in the estimate.

If you and the supplier basically agree on the dollar values for direct labor and material, the $45,000 variance represents a 29-percent increase over estimated costs on manufacturing overhead, general works and administration, and profit. Agreement on direct labor, direct material and manufacturing overhead costs would denote a 73-percent increase over estimates on general works and administration, and profit. Also, concurrence on the amounts for direct labor, direct material, manufacturing overhead, and general works and administration would mean a 220-percent premium on estimated profit.

To offset any tendency to rationalize a variance by adjusting the estimate to match the quoted price, this type of analysis should be performed only after the completion and establishment of a high-level confidence in the estimate.

Savings Possible. Estimating and quote analysis can be extremely useful tools for the procurement organization. Understanding both will add to the buyer's expertise and promote a more professional approach to procurement. Savings are an important byproduct. The cost associated with identifying large premiums and negotiating their reduction or elimination often amounts to less than five percent of the resulting savings.

A relatively small across-the-board savings on procurement costs can significantly increase profits for most organizations, a potential value that is difficult to ignore. Increased profits through improved procurement can be realized by increasing the effectiveness of the buyer with a broader understanding of the value that he is receiving for the price that he is paying. ◀

Summary of Procurement Estimate

Labor and Burden Rate Structure Item Cost	Equivalent Rate per hour	Program Value for 100,000 Pieces
Direct Labor	$2.50	$ 62,500
Direct Material	.34	8,500
Manufacturing Overhead	3.70	92,500
General Works and Administration	1.64	41,000
Profit	.82	20,500
Totals	$9.00	$225,000

How to Simplify Cost Estimates

Here's a simplified approach to the old problem of developing accurate cost estimates. Chief among its numerous advantages is the ability it provides to quickly respond to changing technological and economic conditions

SAMUEL L. YOUNG
Sundstrand Corporation

THE PROBLEM OF COST ESTIMATING in a large manufacturing job shop environment is time-consuming, subject to clerical error, and insufficiently responsive to management's needs for the evaluation of alternatives and rapid change. In general, estimators are slow to respond to management's desire to play the "what if?" game. As an example, what would be the results if the installation of new equipment permitted a five percent improvement in efficiency? Questions of this type are important, for in the competitive manufacturing environment it is often desirable to consider various alternative production and pricing policies. Unfortunately, some of these alternatives are not identified until the basic estimate is known. Then, based on knowledge of the market, management may want to know the impact on total cost if direct labor increases five percent, or manufacturing efficiency improves by ten percent.

In the absence of a well-designed computer program and plenty of available capacity, it is difficult to obtain timely responses to such fundamental considerations. The reasons are found in the tedious methods normally used for the preparation of most manufacturing estimates.

Fortunately, there's a way of getting around this problem. It's a technique based on the use of a summarized historical experience curve utilizing actual-to-standard ratios, and monetary cost-per-hour curves to minimize the number of calculations required. The actual-to-standard ratio relates the number of actual hours required to produce one standard hour.

Summary Curves. The principal difference of this approach, relative to traditional learning curve estimates, is the use of summary curves for both hours and dollars, and actual-to-standard hour ratios. Expressing experience in this form enables the estimator to calculate and plot the data just once. Future estimates can be accommodated by the basic experience curve expressed in this manner. In contrast, the traditional learning curve approach utilizes hours per project and requires each estimate to be separately calculated and plotted, and then extended by a number of financial factors through which hours are converted to dollars.

Use of the proposed summary curves reduces the number of calculations required to a minimum. Once the basic data are calculated, a clerk can estimate any project in a matter of minutes rather than the hours normally required.

The Problem. Typically, preparation of a manufacturing estimate involves several detailed, time-consuming steps. They all start with a definition of the task followed by estimating of standard time requirements and the addition of allowances for nonstandard activities. These include additions for scrap, rework, inspection, learning curves, and the like. The separate addition of these factors is tedious and time-consuming. It also provides numerous opportunities for clerical and computational errors. A typical manufacturing estimate is shown in TABLE 1.

Use of the approach shown in TABLE 1 is further complicated when labor hours are converted into dollars. Each financial factor is separately added to provide the total dollar estimate to be presented to management for approval. Thus the total number of manufacturing hours is multiplied first by the direct labor rate per hour, then by the manufacturing overhead rate, miscellaneous overhead categories, general and administrative and profit factors. This process is illustrated in TABLE 2.

During the management review there may be changes or modifications, hence the estimate may be changed several more times until it meets with management approval. After approval, it is formalized for submission.

It is easy to see that this somewhat cumbersome process does not lend itself to fast, comprehensible change. As a result, it's difficult to quickly evaluate the impact of a change in one or more factors on the net cost.

Another major consideration in the estimating process is the multitude of sources involved in estimate preparation. Typically, several organizations prepare and review an estimate prior to its submittal to management. Staff personnel who prepare or review an estimate may tend to play it safe by adding complexity or safety factors.

Table 1—Manufacturing Estimate for Direct Labor Hours (350 "X" Units)

Standard Hours	100	
+ Setup (Prorated)	27	
Subtotal		127
+ Efficiency Loss	39	
Subtotal		166
+ Scrap Adjustment Rate	54	
Subtotal		220
+ Inspection	41	
Subtotal		261
+ Learning Curve Factor	120	
Total Proposed Hours		381

By the time management receives an estimate, particularly in large organizations, it has passed over the desks of numerous staffers, each of which may have applied his own "feel of the project." As a result, data finally used for the estimate may bear little resemblance to the data prepared by the estimators.

This process is further complicated when management develops the habit of reducing all estimates by some factor, say, 20 percent. This provides staff with one more incentive to play it safe through the addition of contingency factors. To summarize, estimating can be improved by reducing the number of "estimators" involved. This, in turn, requires a significant reduction in the number of calculations used, and increased responsiveness to the evaluation of alternatives.

This proposed approach is feasible only if manufacturing performance is in a reasonable state of control. Performance must be sufficiently consistent to permit plotting of a representative trend. Further, this implies good management information by which nonstandard performance can be separated and isolated. This is not to suggest that variances do not occur in well run plants. Rather, it means that variances and nonstandard conditions must be identified and shown separately for control and management visibility.

Simplified Pricing Preparation. Implementing use of summary curves involves the following steps:

1. Prepare a worksheet as shown in TABLE 3 to determine actual experience by product, project, or summary performance. Company policy may be used in lieu of

Table 2—Manufacturing Cost Estimate for "X" Units

Manufacturing Direct Labor Hours by Direct Rate/Hour at $4.00	381	
Direct Labor Dollars	$1524	
by Manufacturing Overhead at 300%	4572	
Total Manufacturing and Overhead		$6096
by General & Administrative at 10%	610	
Manufacturing Direct Labor, Overhead & General & Administrative		6706
Profit at 20%	1342	
Total Price		$8048

1. CUMULATIVE AVERAGE experience curve. To use this curve, estimator raises a perpendicular from the point indicating required production (in this case 60) to the curve. He then moves left to obtain a value for the actual standard hour ratio (in this case 4:1). This factor multiplied by the estimated labor content per unit multiplied by the number of units required provides a figure of 2400 direct labor hours. ▼

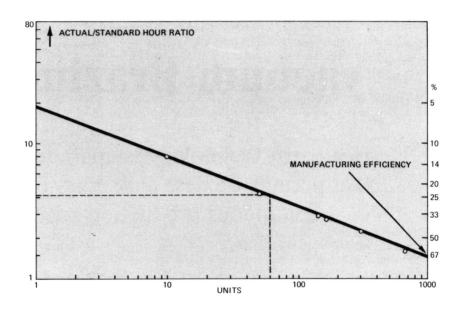

Table 3—Experience Curve Worksheet Actual/Standard Hour Ratio

PRODUCT	UNITS PRODUCED	TOTAL DIRECT HOURS	CUM. AVERAGE ACTUAL HOURS/UNIT	STANDARD TIME (HOURS)	ACTUAL/ STANDARD
1	10	3,900	390	50	7.8
2	50	21,000	420	100	4.2
3	143	79,945	559	190	2.9
4	162	65,480	404	151	2.7
5	305	100,018	328	147	2.2
6	678	155,688	230	141	1.6

Table 4—Cost per Hour of Labor

Direct Labor Dollars	$1542
+ Manufacturing Overhead	4572
+ General and Administrative	610
+ Profit (Optional)	1342
Total Price	$8048

Total Price ÷ Direct Labor Hours = Price/Hour of Labor.
$8,048 ÷ 381 Direct Hours = $21.12/Hour of Labor.
To Complete the Estimate:
   1. Total Direct Labor Hours = 2400
   2. Price per Hour = $21.12
   3. Multiply (1) x (2) = $50,688.00

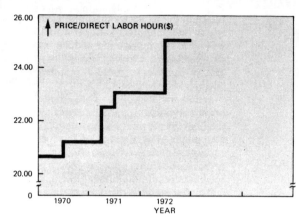

2. TYPICAL GRAPH showing price of direct labor on an hourly basis. Figure of $21.12 obtained from this graph multiplied by 2400 direct labor hours (from example in Figure 1) gives a total direct labor cost of $50,688.

actual experience

2. Using this worksheet, determine the number of actual hours required to produce one standard hour of effort. This is called the actual-to-standard-hour ratio

3. Plot the actual-to-standard-hour ratio on log-log graph paper in the appropriate unit position as shown in *Figure* 1. For example, if at the 50th unit the actual-to-standard ratio is 4.2, plot that value at the 50th unit

4. Plot enough data to determine the characteristics of the data and simplify curve fitting. Be certain that the data used are representative of actual performance

5. Fit a curve to the plotted points. This is your experience curve and it will be the basis for preparing all future estimates.

Simplified Estimating. Once the experience data are calculated, the estimating process is quite simple. The procedure for estimating the labor hours required in any project is as follows:

▶ Define the product and determine its standard hour content.
▶ Determine the number of units to be produced.
▶ Refer to your experience curve (*Figure* 1). Locate the number of units to be produced on the horizontal axis.
▶ Follow the unit perpendicularly upward until it intersects the experience curve.
▶ Follow the curve intersection to the left until it intersects the vertical axis.
▶ Identify the actual-to-standard ratio at the point of intersection. The comparable manufacturing efficiency is shown opposite on the vertical scale to accommodate alterations or complexity judgments.

This procedure can be illustrated by assuming an instance in which it is necessary to estimate the number of labor hours required for 60 "X" units with an estimated labor content of 10 standard hours. The procedure is as follows:

1. Find the 60th unit on the horizontal axis of *Figure* 1
2. Follow up to intersection with the experience curve
3. Follow the point of intersection to the left until the line thus produced intersects the vertical axis
4. At the intersection, note the Actual/Standard ratio of 4:1. Your experience indicates that to produce 60 units you will need an average of four actual hours to produce one standard hour
5. Multiply the "X" unit standard hour content of 10 hours by the 4:1 actual-to-standard ratio. This results in an average of 40 hours per unit. The average efficiency would be 25 percent (1 earned divided by 4 actual hours)
6. Multiply the 40 hours per "X" unit by the 60 units resulting in a total of 2400 labor hours (40 hours/unit multiplied by 60 units).

Conversion of labor hours to dollars requires multiplying by the cost or price per hour of labor derived in TABLE 4. If your projects take place over a period of years, "graph" the cost or price per hour data including likely price adjustment factors. This should reflect known and estimated adjustments as union agreements, plant load, market conditions, *etc.* Figure 2 permits extraction of the properly adjusted price directly from the graph for the year or years concerned. Once prepared, the graph suffices until some fundamental change occurs. Added to this, of course, will be the cost of material and purchased services.

Admittedly, the use of graphics introduces the possibility of small errors in plotting and extrapolating. In the long run, however, the advantages of speed and simplicity far outweigh these small disadvantages.

To summarize, the proposed approach dramatically reduces the number of required steps and calculations. Further, the reduction in calculations minimizes the possibility of clerical errors. As a result, the estimating process meets its original objective—greater speed and more effective response to the needs of modern management. ◀

Variable Cost Estimating

Phillip F. Ostwald
College of Engineering and Applied Science
University of Colorado

Gary E. White
College of Business
University of Colorado
Boulder, Colorado

Abstract

Variable cost estimating is a superior method useful for product costing, cost control, profit planning, and pricing. This paper discusses estimating procedures for labor, material, tooling, and variable overhead. Applications for product cost, design-to-cost, and make or buy are covered.

Introduction

Variable costs are defined as costs of raw material, direct labor, and direct supplies plus other predetermined and identified indirect costs which vary closely with production quantity. It is agreed beforehand that these indirect costs can be reasonably allocated to a product or a line of products. Estimating these variable costs is handled primarily by industrial engineers. In view of the concentration on this kind of estimating, it is titled variable cost estimating.

Analog terms are used by other professionals. The accountant uses "direct costing" while the economist likes "marginal costs". Another term, sometimes used improperly in certain contexts, is "out-of-pocket costs". While we do not wish to engage in semantics, variable cost estimating more aptly describes the act of estimating these costs than do other terms.

There are some 300,000 U.S. firms who manufacture a product. The majority of these firms use the operation or department method[1,2] of product estimating. In our opinion variable cost estimating is superior. This paper summarizes the method, and challenges the reader to compare and judge the advantages and contemplate the problems in switching to variable cost estimating.

Initially we describe the history of variable cost estimating. Inasmuch as information is always required, a flow chart illustrates key information necessary for this kind of estimating. Overhead and other indirect costs must be separated into variable and fixed components. Once this cost constructing is achieved, products may be estimated. Other applications are stated and finally limitations which impede the introduction of the technique are given.

History of Variable Cost Estimating

Viewed in a historical perspective, variable cost estimating has no specific date of origin. Earlier techniques arose as accounting developed from its bookkeeping role in preparing financial statements from accounts. This histroy is long and obscure, but as a need for accurate product data emerged out of the industrial revolution, it was natural that variations in estimating and cost accounting techniques were attempted to satisfy the growing management and owner sophistication. As a new company, plant, or ownership developed or altered, the variety of techniques matched the ingenuity of owners and managers and their objectives. Approaches to communicate product costing and cost-volume-profit relationships proliferated. One thing can be said with certainty, variable cost estimating is not a recent development or a unique kind of cost system, though its formalization probably was in the thirties. It is a feature introduced into a cost system and cost estimating to broaden usefulness of the information.

Early development of accounting theory emphasized and required the use of absorption or full costing techniques, i.e., the application of both variable and fixed manufacturing costs to the carrying values of manufactured products. Tax laws accommodated to the accounting approach of reporting inventory values, although Internal Revenue Service interpretations have apparently not been totally restrictive in this area until only recently.

As technology prospered the role of the master mechanic unfolded into the industrial engineer. Early applications in time study, production methods, and shop management gave him a special awareness or shop costs. At first special studies were undertaken to meet various management requests for lower shop costs and the like. Inasmuch as accounting data were historical these special studies were separate from that kind of information. Standard time data, evolving from time study, improved the success in estimating future labor costs. Simple blue-print reading skills, traditionally taught in engineering programs, expanded into a complex ability unmatched by other disciplines. Intracacies in the configurations of materials were determined from the drawings designed by other engineers. Long time associates with time studies, material take-offs, and methods and process engineering were requisite knowledge for operation and product estimating, a specialization within industrial engineering.

In the more recent decades, advances in techniques established the need for a different emphasis in data. Operations research was notable in these efforts. As a consequence the association between engineers and accountants have strengthened and the examination of the origin, preparation, and use of cost data are receiving emphasis. Variable cost estimating is one result, and the concentration upon its identification and exploitation has established its place as a technique for internal cost analysis and cost estimating.

Flow Chart for Cost Information

Costs are desired for a variety of purposes. Before one can construct an idealization of information flow, these purposes must be determined. Insofar as the cost estimating function is concerned, the first priority is to estimate costs to obtain new product business and retain old product business. Other internal purposes such as incentives, cost control, value engineering, project analysis, etc., while important, are secondary. However, primary and secondary uses must be coordinated to meet all needs with relevant figures. Estimating procedures are often independent of financial accounts and statements and that view is represented here. Though this is a parochial estimating stance, it simplifies an orderly listing of information to undertake variable cost estimating for product designs. Figure 1 itemizes policy, historical, and measured data, and their likely sources. It should be noted that the data and the ultimate variable cost estimate are incomplete for purposes like pricing, engineering change orders, delivery lead times, and profit planning.

Separating Overhead

In absorption costing the variable and fixed indirect costs are mixed in the total overhead rate. This blending obscures the theoretical and practical differences. In variable costing fixed costs are expensed in the period incurred, but absorption costing applies some of this fixed cost to unsold inventory through the total overhead rate. First, we show conceptual differences by manipulation of simply functions[5]. Profit under variable and absorption cost are given as

$$P_{VC} = (Q_S)SP - Q_S(M+L+VOH) - FOH \qquad (1)$$

$$P_{AC} = (Q_S)SP - Q_S(M+L+VOH)$$

$$-Q_S\left(\frac{FOH}{Q_N}\right) - (Q_N - Q_P)\frac{FOH}{Q_N} \qquad (2)$$

where SP=unit selling price, M=material costs per unit, L=labor costs per unit, VOH= variable overhead per unit, FOH=total fixed overhead for the period, Q_N=normal capacity, Q_P=actual quantity produced, and Q_S=quantity sold. Fixed costs charged against revenue in the two systems are variable costing: FOH (3)

and in absorption costing:

$$Q_S\left(\frac{FOH}{Q_N}\right) + (Q_N - Q_P)\frac{FOH}{Q_N} \qquad (4)$$

The first term in (4) is the standard fixed overhead and the second is volume variance, both of which use the standard fixed burden rate FOH/Q_N. If $Q_S \neq Q_P$ the distinction between the two systems is the difference between the fixed costs that are expensed, or equation (3) minus equation (4) or

$$(Q_P - Q_S)\frac{FOH}{Q_N}. \qquad (5)$$

Setting $Q_S = Q_P$ in (4) it is seen that the fixed costs charged against revenue are identical under both systems, or

$$Q_P\left(\frac{FOH}{Q_N}\right) - (Q_N - Q_P)\frac{FOH}{Q_N} = FOH \qquad (6)$$

For the predetermined overhead rate, as given by the term $\frac{FOH}{Q_N}$, the volume variance in absorption costing is measured by

$$(Q_N - Q_P)\frac{FOH}{Q_N}. \qquad (7)$$

The next question becomes one of making a practical division between FOH and VOH. Indirect cost accounts for a plant or department are analyzed as to their character. Does the cost vary or is it independent of quantity changes? The response to this question is affected by circumstances which vary from company to company and from time to time within a particular company. Even after a classification is made, some costs are shifted from one category to another. Objectives should indicate the category of costs. Where profit planning objectives predominate, borderline indirect cost items are usually specified in the variable cost category to avoid unprofitable decisions by the understatement of direct costs. Another emphasis is in cost control which leads to the placing of certain costs in fixed overhead - those costs not fully controllable by supervisors or plant managers. Cost classification into variable or period categories is a step in preparing the budget for some future period. Usually the chart of accounts is examined via a conference and open discussion. Systematic study and analysis of each machine or department at different levels of production allow curve plotting, etc., to see if a correlation is possible. Often times engineers and accountants jointly participate in these studies. Table I examines one study, necessarily simplified in detail, and it shows for one company the division of overhead costs. Standard books in accounting and reference I may be examined.

In Table I, the $164,216 of fixed overhead is covered by contribution to variable costs. The $317,422 of variable overhead is then applied as a percent to base labor. For example, if for an identical period the base labor dollars was estimated to be $185,000, then the variable overhead percentage is (317,422/185,000) x 100 = 171.6%. This rate then multiplies the direct labor dollar estimate to give the corresponding variable overhead.

The Product Estimate

The product estimate document ranks foremost in the economic evaluation of a product. Viewed correctly, the product estimate is a report and is composed of many separate estimates such as labor, material, and tooling. A form, often called a "recap", is the summarization of the report. This recap is input information to the price setting function.

Considering the number of firms that produce a product, variety in variable cost estimating can be expected, and our purpose is to discuss a few of these practices. In the following three estimates and one pricing computation, headings and terms will vary, and these forms could be expanded or consolidated in diverse ways. These estimates may be either preliminary or detailed and the experience and judgment of the estimator is critical in their findings. For instance, if material is the most significant cause of cost, then it would be natural that it receive major attention.

Table 2 is a typical material estimating form. Organization of the document primarily follows bill of material part numbers for several reasons. The bill is the fundamental authority and is a complete and correct listing. Thus no materials will be overlooked. The buildup of materials is from raw materials, standard purchased parts, manufactured components, and minor and major subassemblies to the packaged product ready for sale. Thus, a subassembly material cost will be the sum of materials that unfolds into it. False entry of other material costs will not be inadvertent. Furthermore, as one product may call for a certain interchangeable subassembly, it is important that the part-numbering organization permit this material estimating flexibility.

The material estimate considers three sources of direct material: raw material, standard purchased material, and manufactured material. The distinctions are major ones in estimating. Raw material, say stored in general inventory, is processed on while standard purchased material is used in the condition as received. Tires for a car are a typical example. Manufactured material, meaning raw material internally worked on, may be used for several different products. The source column in Table 2 could be a numeric code adapted to material requisition procedures such as standard hardware, vendor F make, plant B make, division A or C make, etc.

The Unit/Product column calls for a material quantity or volume take-off using design drawings. Either estimating or design engineering is charged with material take-off. The Price/Unit may be estimated or quoted and usually purchasing or estimating is responsible for obtaining the cost of the material. Price breaks due to quantity discount would be noted in this column. Scrap and rework, returned material values, runners, splits, credits for byproduct materials, multiple-unit materials, etc., call for a special material estimating understanding that is not discussed here.

A variable overhead rate can be specifically applied to material. It would include those indirect cost elements attributable to material handling, receiving inspection, FOB, dock and inventory costs and losses, etc.

The labor estimate, see Table 3, is organized along lines of the operations process sheet. The estimator sometimes processes the part even before it is assigned to manufacturing process specialists. Naturally this practice varies depending if the estimate is preliminary or detailed and on the firm. In our simplified example, the operations-processing part of the form has been suppressed.

In variable cost estimating, an essential step is to tie the labor either to a department or machine. If this is done, the variable overhead portion can be structured similarly for greater accuracy.

The processing plan is considered stable despite nominal changes in lot quantity. It would be expected to change for greater differences in quantity. The extended cost per unit is found using

$$\text{Cost per unit} = (\frac{SU}{Q} + \text{Run Hrs})(\text{Labor Cost/Hr}) \qquad (8)$$

where consistent dimensions are implied. This estimating approach is keyed to lot release rather than total sales forecast, annual production, and the like. For job shop estimating setup is a major cost. For long production runs the approach is less valid as $SU/Q - 0$. Under these circumstances setup cost may be placed in the variable overhead portion and ignored in equation (8). In Table 2 three lot sizes are considered and one will be eventually recommended. It is important to recognize that the labor cost is not a constant per unit. While many estimators treat it as a constant, we recommend that procedures be incorporated to allow for varying labor cost per unit.

The "Setup" and "Hours per 100 Operation" are standard, i.e., allowances have converted normal time. These are the usual practices. Hours per 100 units, minutes per piece, and units per hours and others are equally acceptable and widely used. Performance against these standards for either incentive or day work is reflected in the variable overhead rate by a predetermined or historical variance amount.

Operation 10, "mold 6 per shot," is cost-calculated on a unit shot basis, i.e., labor input divided by six.

Portions of the work content for a new product may be identical or similar to the work content of segments of existing products. Standard or actual costs could be used in place of estimates. It is necessary that these substituted costs reflect the conditions under which the manufacturing will be carried on.

The Tooling Estimate Log, as shown by Table 4, is the summary of the tools for this product. A tooling estimate, not shown in this paper, indicates material, labor, tooling overhead, and perhaps a percentage tooling contingency. Each tool, exclusive of perishable and standard tools, as indicated by the operations process sheet is estimated. The estimate is often made by an experienced tool estimator without the advantage of a formal tool design. Standard time data for labor, material requirements and costs, and tooling overhead are calculated. The tooling overhead includes provision for tooling design, supervision, space, depreciation, etc. The estimate reflects the cost of the tooling as if the tooling department operated independently of the manufacture of the product. The make-or-buy decision is made more objectively if tool cost is estimated this way.

The policy of the tool write-off is important to the cost estimate. One of two policies is followed: 1) charge cost to specific customer who receives title but may or may not house the tool following use (other variations with ownership exchange rights are possible) and 2) prorate the tool over so many units, lots, or years. If a conservative policy is possible, management is interested in recovering tooling costs as quickly as possible because of the fears of obsolescence, part changes, and overstated sales. But pricing competition and rules by some large companies and the government may not allow write-off on the initial lot. In some respects the tooling write-off is a hidden pricing decision. Immediate write-off increases cost and causes higher prices while deferred amotization keeps recovered tool costs minimal. Insofar as variable cost estimating is concerned, the proration of tooling costs is consistent with a variable consumption of value as volume varies. If the tooling is to be amortized over X number of units, later estimates will be less by this amount. Maintenance of tooling is a consideration then.

The Unit Pricing Computation form has many variants. The one shown by Table 5 does not reflect the possible detail and computation. This one, analyzed for a lot quantity of 122, would require computations for other lot quantities. The cost estimate is perhaps the most critical input information to this document. Usually price is set by the executive or accounting or profit planning group along with cost estimating.

Extended labor cost per unit is multiplied by the variable overhead percentage. Material, labor, and variable overhead are summed. Certain out-of-pocket costs are separately estimated. These items furnish additional variable cost.

Another variation to tooling and setup costs is to deduct the entire amount from net variable profit dollars. In this way, they are treated as fixed costs. The deduction for other period or fixed costs is also made at this point. The remainder is now profit. This is the usual contribution approach. While product estimating using varible cost estimating is the major application, we now direct our attention to instances where this technique can help other programs.

Design-to-Cost

For the most part, product estimators build up an estimate starting with operation estimates and concluding with either product cost or price. Sometimes a reverse procedure is required. Beginning with a competitor's market price or a management price objective, the estimator works backward to find total cost and the cost for various design elements. This practice is called design-to-cost. Profit, General and Administrative, Selling, and Engineering costs are taken out of the price leaving Cost of Goods Manufactured.

The apportioning of the Cost of Goods Manufactured is in accord with a logical design structure for the product. Designers consult with the product estimator in developing the logical structure. Each design element is given a design-to-cost "target." The designer knows that he controls his design and hopefully is able to match the goal. Figure 2 is an example of the distribution of a $555 priced object.

The designer is given the variable cost goal. In Figure 2 the Cost of Goods Manufactured of $415 is broken down into Fixed Overhead of $195 with the $220 remainder representing variable cost. It is the $220 that is estimated and associated with design-to-cost goals. Furthermore, the Power Supply with a "design-to-cost $25" could be broken down in variable costs for the battery, relay, and regulator if desired. It is additionally possible to break down the variable cost into labor, material and variable overhead if desired although this may become too restrictive in a practical design sense.

Engineering change orders, a frequent problem with some kinds of products, can be evaluated with variable cost estimating. The production cost per unit increase or decrease for direct materials, direct labor, variable overhead, and tooling are the principal economic factors calling for the approval decision.

Make or Buy

Short-run profit planning considers production runs or units which are within the existing capacity or time duration which prevents major changes in capacity. Total period cost remains constant in these circumstances. The make-or-buy decision with parts, subassemblies, etc., using variable cost estimating can be made within this context.

If a company has unfilled capacity, the additional cost of making parts consists of variable costs. The cost to be compared with prices quoted by suppliers should be the amount of cost to make which will not be incurred if the item is purchased. This cost will vary with in-house activity. During periods of low activity, capacity may be available for additional work at low out-of-pocket cost. In periods of high activity it may be necessary to work overtime or acquire more equipment before additional work can be scheduled.

Where short-run policy is to retain employees during periods of low business some of the direct labor costs may be overlooked. For instance, a toolmaker performing production work during times when they would be otherwise idle.

Cost Control and Other Advantages

Separation of costs into variable and fixed categories is the first step in improving short-run cost control. Variable costs are generally "out-of-pocket" costs and are controllable in terms of the prices paid for the respective cost factors and also in the efficiency in using these inputs. Efficiency measures must be tied to volume levels and estimated variability of the costs. Fixed or period costs are generally not controllable in the short run with respect to efficiency. Within a range of attainable volume levels any fluctuation in the input/output ratio has little influence on the amount of fixed cost incurred. Prices of certain fixed cost factors may change from budgeted levels, but in most cases a heavy portion of the fixed costs have been committed in prior periods and are currently absorbed as capacity charges, i.e., depreciation.

A list of advantages for variable costing includes the following:

1. Changes in inventories do not affect periodic profits because fixed costs are expensed when incurred.
2. The true nature of fixed costs as "period costs" is emphasized by presenting them in total.
3. Profits increase and decrease only as sales increase and decrease.

For decision-making and control:

1. Separation of costs into fixed and variable categories improves control.
2. Costs controllable at operating levels can be separated from those controllable by only middle or top management because deceptive cost allocations are avoided.
3. Flexible budgets and standard costs are relatively easy to develop.
4. The marginal contribution of product lines is readily available.

5. Data required for cost-volume-profit analysis are more readily available in accounting systems using variable costing than in those using absorption costing.

The primary disadvantages of the technique include its general unacceptability for external reporting, and the potential danger that in the long run variable costing may result in under-priced production since the product does not include fixed costs.

External Reporting

Variable costing has not been formally accepted by the American Institute of Certified Public Accountants in certifying externally reported financial statements, although it has been used by some firms and many accountants propose that it should be formally accepted. One source[4] states:

> "The primary basis of accounting for inventories is cost, which has been defined generally as the price paid or consideration given to acquire an asset. As applied to inventories, cost means in principle the sum of the applicable expenditures and changes directly or indirectly incurred in bringing an article to its existing condition and location.
>
> It should ... be recognized that the exclusion of all overheads from inventory costs does not consititute an accepted accounting procedure."

The Securities and Exchange Commission takes a similar stance for financial reporting which falls within its province.

Probably the most potent and pervasive force in support of absorption costing is the Internal Revenue Service. In 1973 the IRS issued regulations (Reg. 1.471-11) for inventory costing which made the absorption method mandatory. Prime costing and direct costing methods were specifically outlawed, but the complex rules did permit the continued use of direct costing in some limited cases. The IRS favors absorption costing because it forces the firm to defer the deduction of fixed factory overhead from taxable income until the period in which the product is sold.

Since many firms realize advantages by using variable cost estimating internally, adjustments must be made to the internal data to provide external reports on the absorption cost basis. Comparatively simple procedures determine the amount of the periodic adjustment. Note that the difference in net income between the two methods is equal to the difference between the beginning and ending inventory in units times the unit cost of the fixed factory overhead applied with the absorption method. This is shown by equation 5 as the difference in net income arising from different production and sales quantities in the same period.

Summary

This paper has discussed variable cost estimating. On going cost estimating would benefit with a variable cost system compared to an absorption overhead system. These reasons, short-run cost control, profit planning, and production, are advantages given by most companies who use variable cost estimating.

References

1. Ostwald, Phillip F., *Cost Estimating for Engineering and Management*, Prentice-Hall, Inc., Englewood Cliffs, NY, 1974.

2. Ostwald, Phillip F. and Gary E. White, "Product Cost Estimating," *AIIE Technical Papers*, 1973 pp. 105-116

3. DeCoster, Don T., and Kavasseri V. Ramanathan, *The Accounting Review*, Vol. XLVIII, No. 4, Oct. 1973, pp. 800-801

4. *Accounting Research and Terminology Bulletin*, Final Ed., Accounting Research Bulletin 43, pp. 28-29, AICPA, 666 Fifth Ave., New York, NY 1961.

5. *Current Applications of Direct Costing*, National Association of Accountants, NAA Research Report 37, New York, NY 1961.

	Total Cost	Fixed	Variable
Indirect Labor			
Supervision	18,340	17,257	1,083
Inspection	22,470	1,905	20,565
Receiving	2,108		2,108
Maintenance	45,117	3,704	41,413
Watchman	2,288	2,089	210
Setup	26,807		26,807
Toolroom	20,165	2,075	18,093
Inventory	51,348	4,117	47,231
Vacation package	31,417	31,417	
Time study	898	898	
Tool engineering	5,904	5,289	615
Planning	2,702	2,442	260
Payroll	6,850	4,268	2,582
Factory clerks	10,163	9,565	598
Engr. product development	4,286	3,606	680
General office	4,249	3,056	1,193
Factory pension	9,569	9,569	
	264,681	101,243	163,438
Manufacturing Expenses			
Perishable tools, dies	26,780		26,780
Production supplies	23,686		23,686
Factory supplies	19,450		19,450
Machinery repair, equipment	25,928		25,928
Inbound freight	3,695		3,695
Factory travel and personnel	1,858		1,858
Machine rental	1,523	1,437	86
Light, power, water, gas	19,525	560	18,965
Outside storage and inventory	5,719	777	4,942
Depreciation - building	5,689	5,689	
Depreciation - equipment	19,975	19,975	
Taxes - unemployment	7,536	7,536	
Taxes - FICA	21,821	21,821	
Taxes - property	3,193	3,193	
Taxes - sales and use	3,078		3,078
Insurance - property	1,985	1,985	
Insurance	25,516		25,516
Total Manufacturing Expense	216,957	62,973	153,984
Total Indirect Labor	274,681	101,243	163,438
Total Overhead	$ 481,638	$ 164,216	$ 317,422

Table 1. Overhead Distribution into Fixed and Variable Categories

Material Estimate

Part Number 902--176 Product Number 902 Date 5/24
Component ☐ Product Description _____ Estimator PFO
Subassembly ☒ Quantity: First year _____
                        Second year _____
Used in P/N 902-0213   Third year _____

Number Specification	Description	Source	Unit of Measure	Unit/ Product	Price/ Unit	Cost/ Product
6682	Masonite board	Inventory	Each	1.000	$0.170	$0.17000
6137-001	Shakeproof	Inventory	Each	10.000	0.00737	0.07370
T202-Dow	Adhesive	Purchased	Gallon	0.00716	2.70738	0.01938
Staples	7/16" x 3/16"	Inventory	Each	29.	0.00036	0.01044
902-0032	Tie strap	Manufactured	Piece	1.000	0.0242	0.02420
1018-313	Elastic 1/8" round	Inventory	Yard	4.235	0.235	0.99523

Total $1.29295

Table 2. Typical Material Estimating Form

Labor Estimate

Part Number 902-0136 Product Number 902 Date 5/24
Component ☒ Product Description _____ Estimator PFO
Subassembly ☐ Lot size 122, 507, 1318
Used in P/N 902-0137 Materials: Mold compound, insert

Dept. & Machine	Operation	Operation Description	Job Class	Labor Cost/Hr	Hrs. Setup	Hrs. per 100 Opn.	Ext. Cost/Unit Q=122	Q=507	Q=1318
803-03	10	Mold 6 per shot	4	6.25	2.25	.725	.12282	.03529	.01822
803-02	15	Hand trim & inspect	6	4.55	.34	1.638	.08721	.07758	.07570
652-17	20	Cutoff ends of insert	7	4.25	.95	1.826	.11070	.08556	.08067

Table 3. Abbreviated Operations Process Sheet Used to Estimate Labor Cost

Tooling Estimate Summary

Product Number   902
Date   5/24
Estimator   GEW

☐ Charge to customer
☐ Include in unit price
☐ Prorate over _____ units
               _____ years

Component Part Number	Tool Part Number	Description	Status	Source	Cost
-0136	902-7010	Injection mold	New	Make	$ 800
-0136	902-7011	Cupping fixture	New	Make	120
-0137	902-6089	Drill fixture	Existing	--	-
-0138	902-7012	Plate	New	Buy	195
----					---
				Total	---

Table 4. Log of Tools Specified by Operations Process Sheet for a Specific Product

Unit Pricing Computation

Product Number   902
Product Name   _____
Lot quantity   122

Basis for prices
Distribution % range   _____
OEM
Other   _____

Dept.	Extended Labor Cost/Unit	Variable Overhead Rate, %	Variable Overhead Applied	Unit Labor + Variable Overhead
01	1.032	131	1.352	$ 2.384
02	4.210	121	5.094	9.304
--				
17	1.822	97	1.765	3.587
			Total	$23.041
			Material per Estimate	4.063

Total labor, material, variable overhead $27.104

Out-of-pocket costs
    Tools and fixtures
    Equipment
    Outside design
    Marketing
    Other

Total out-of-pocket charges ÷ y quantities = 17.210
Total $40.251

Selling price	$60	$62	$67
Quantity sold	15,000	15,000	15,000
Three year minimum sales	90,000	930,000	1,005,000
Less total variable costs	603,765	603,765	603,765
Net variable profit dollars	296,235	326,235	401,235
Net variable profit %	33	35	40

Table 5. Abbreviated Pricing Form for Variable Costing Method

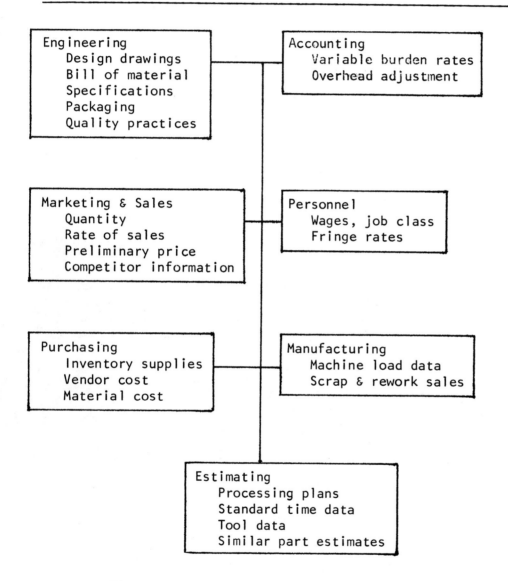

Figure 1. Flow of policy, historical, and measured data to cost estimating

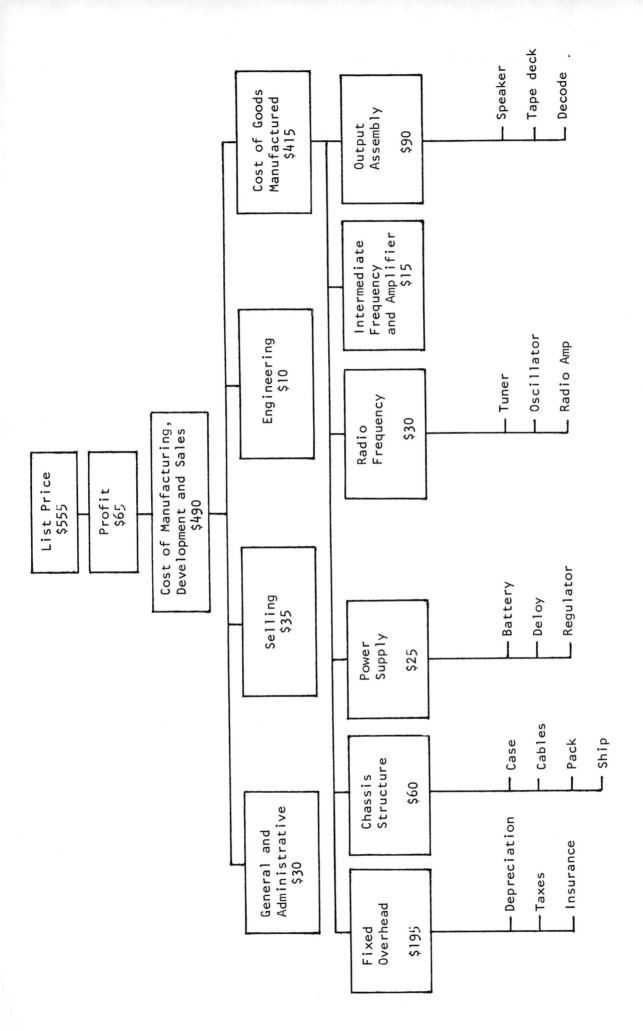

Figure 2. Distribution of price to Design-to-Cost goals for a product

Preparing Estimates for Developmental Work

By Paul D. Fowler
Professor
Georgia Southern College

Major estimates, particularly for developmental or low volume work, should be built up from highly detailed estimates in order to establish basic parameters for manpower skill levels, for facility requirements and scheduling, and for assuring that major cost components of the estimate have not been overlooked. However, a too rigid adherence to the details of the estimate may cause losing sight of broader and even more significant criteria, such as the overall complexity or closeness of tolerance of the product as affecting costs, or of considerations in managing the manufacture of the product to minimize costs.

INTRODUCTION

The various steps required in preparing a quotation are examined with the objective of increasing their effectiveness in not only developing a realistic price estimate, but also in laying the groundwork for managing the production of the product in a more efficient manner. A two phase approach to estimating is suggested in which individual components of the product, or of functional contributions to the product, are first estimated in what is termed a "Details Developed Quote," where the individual component costs are developed in "the most extensive level of detail practical." The quote is then subjected to a comparison with historical data from similar products as a check on its realism, and adjustments are made as indicated. It is suggested that certain types of non-repetitive functions, such as design, be evaluated by "Experienced Based Estimates" only, since estimates of detail tasks give an appearance of "false precision" which really does not exist, and are actually misleading in the same sense as "insignificant figures" in mathematics.

Top management is particularly encouraged to be systematic in developing and providing guidelines to the people making the quote, and in evaluating the completed quote for realism and internal consistency. A caution against arbitrary padding and cutting of quotes is made to quoters and management.

The quoter is urged to carefully document all assumptions for his part of the product quote, including those pertaining to the exact nature of the product(s) to be delivered to the customer, and is urged to share these with his fellow quoters to assure their compliance, and with those coordinating and screening the quote to facilitate their work in completing the quote. His assumptions also can enable the eventual manager of the product to better fulfill his responsibilities in the areas of establishing accountability, of recording and controlling costs, and of diagnosing and correcting recurrent problems in the program.

A TYPICAL QUOTE PREPARATION

A quote typically will be developed either for an internally developed product or service that the company hopes to market, or else for a product or service that another company or organization considers contracting with you to produce. The procedures for quoting your own product are generally an abbreviated version of those for bidding on someone else's product, as follows and as shown diagrammatically on Figure 1. (Figure 1 shown on following page)
    1. Your company becomes aware of someone else's need for a product or service. This may come about through intelligence gathered by your sales people, through advertised re-

FIGURE 1

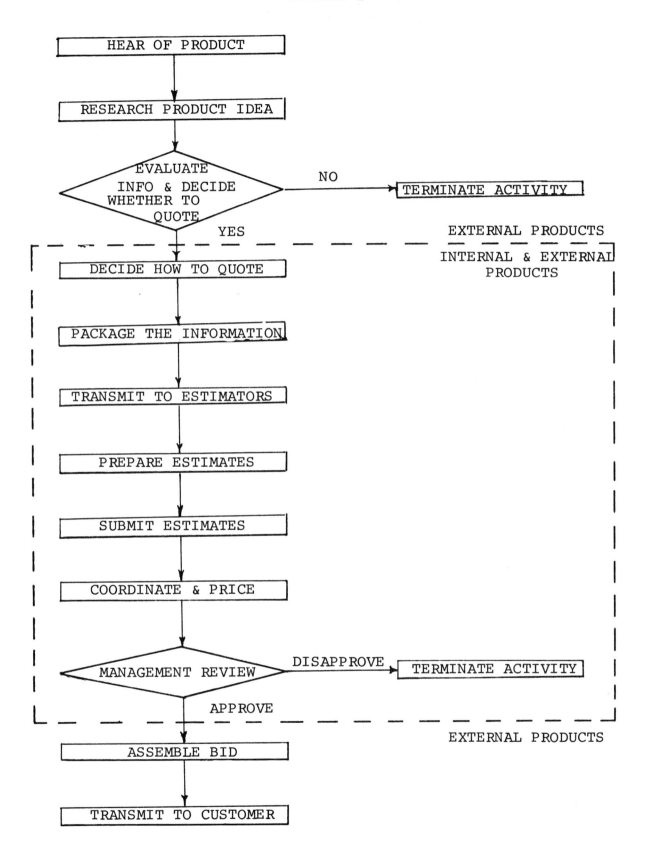

quests-to-bid printed in your trade journals, or through bid packages sent directly to you. Perhaps most commonly for some of you, a representative of one of the companies for which you act as supplier gives you a telephone call to see if you are interested in bidding on a certain item.

2. Next, you have to establish the specific requirements of the need, by requesting drawings, specifications, samples, statements of work, or whatever else you can identify that allows you to determine just exactly what it is that the customer wants. You go about gathering all of this information at a bidders conference or technical briefing or simply by calling on the customer and getting whatever information he can give you. Other times, he may be able to mail you most of the information. In the construction trades, for publicly advertised bids, the specifics may be available only at a particular location, and you may have to secure your information on the spot.

3. Once all of the available information from external sources is in hand, a determination must be made as to whether a quote will then be made and a bid prepared. The considerations in this decision, treated in detail later in the paper, generally deal with whether the company is sufficiently endowed with the necessary technical expertise, equipment, manpower, and space as to have an expectation of producing at a profit. As will be seen, this final decision is one requiring a top management decision. This step in the general procedure would be the first if the company is considering producing its own design, in which case all of the technical information regarding the product should already be available.

4. Next, those responsible for coordinating bids in the company must determine how to get the necessary information to all those who will contribute to the bid preparation. This can be quite an involved undertaking in a large company, such as is common in the aerospace industry.

5. Packages of information must be put together for those who will make the estimates. This can involve preparation of preliminary designs, engineering work statements, schedules and other data.

6. Finally, a meeting may be required to distribute the information to all those who will quote, and to establish mutually agreeable assumptions for aspects of the product requiring coordination among the various participants in the company. Examples are completion dates for the quote and specifications pertaining to the product itself, such as finishes, technical data requirements, etc.

7. The estimators develop the quotes.

8. By the established due date, persons preparing individual portions of the quote, such as those for outside procurement, tooling, engineering, etc, must submit them for coordination. Coordination can take place in different areas of the company, since it is primarily the determination that the program assumptions laid out by the individuals making quotes do not conflict, and that the relative sizes of the various parts of the quote indicate the right proportions to one another.

9. Next, dollars and cents prices must be developed from the foregoing estimates, which were expressed in a combination of units, such as prices, various types of manhours, facility requirements, etc. Allowances will be made for the learning curve effects for long runs of very complex products, for wage and price escalation, for variations in the quantity of products ordered and for other factors.

10. Management must then review the resulting estimates and assumptions, making compari-

sons with cost performance on similar programs, and intelligence concerning prices of the competition, and then decide whether to submit the bid as it is. Several cautions pertaining to this step will be discussed later in the paper. This is the last step for quoting an internally developed design.

11. The quote package with prices, assumptions, and technical descriptions will then be assembled into the correct form for transmittal to the customer. Many things can go wrong at this step, so some companies employ persons who specialize in editing, printing and otherwise assuring the accuracy of the finished bid package.

12. Finally, the bid package is transmitted to the customer, probably by courier or by registered mail.

Now that we have described the usual procedures for developing information for the quote, for getting it out to the necessary people, and for reviewing and pricing it, we need to look in more detail at the ways that the individual estimates themselves are developed.

SEPARATELY QUOTED FUNCTIONS/COMPONENTS

Thus far, we have indicated that several people may be involved in developing parts of the estimate, but we have not described these parts. For a very complex product, there may indeed be several parts of the estimate, broken down into component part of the product, or perhaps by the organization unit of the company that will contribute to its manufacture. More likely, there may be a combination of the two approaches, as where separate estimates are developed for the engineering, tooling and manufacturing functions, and still a separate estimate is solicited from an outside source for some components that are beyond the capability or desire of the company to produce independently. The cautions given in the preceding list of steps concerned such matters as assuring that engineering is not assuming an ultra-sophisticated manufacturing technology in order to simplify its design effort. I grant you that this will not happen if your people are accustomed to talking with one another, but my experience is that they many times do not. Anyway, those coordinating the estimate should be alert to any such mismatch in the parts of the estimate that results from using conflicting assumptions.

Another of the necessities for producing a good quote, with all of the parts based on the same kind of program, is a program schedule.

DEVELOPMENT AND USE OF A SCHEDULE IN QUOTING

The importance of having a reasonably accurate program schedule for use in making the estimates can hardly be overemphasized, since it determines the time allotted to the various program tasks such as design, tooling-up, fabrication, assembly, and testing, thereby imposing secondary constraints on the quote such as the number of people and machines that must be employed to do the job within the schedule constraints. Typically, the schedule will be developed by backing off from the required completion date, if it is mandatory, or else by building forward from the go-ahead date if it is not mandatory. Naturally a lot of negotiation and adjustment is required to put together such a schedule, making it a fairly high-level task which is beyond the scope of this paper.

Another necessity for making a good quote is a set of assumptions and also historical data relating to similar products.

DEVELOPMENT AND USE OF QUOTING GUIDELINES AND COMPARISONS

It is essential that a top management person, or someone with a high level perspective of company operations, the markets for its products, the company financial picture and the overall state of the economy, both present and future, put together, for quoters, a comprehensive set of guidelines for making their estimates. My experience in working in an estimating capacity at a fairly low level in the organizational hierarchy indicates that management assumed either that I did not need such guidelines, or that such information was common knowledge. I rather think that they believed the former to be true, which certainly didn't help my pride in the job, or my motivation for taking into account all the necessary factors in making a quote.

Some of the information that should be included in such guidelines would include the following:

1. Assumptions as to the degree of likelihood of other new products and activities being also in progress during the period of time in question, including their production rates and requirements for the same space, facilities, and manpower that are required for the product in question.

2. Identification of the competition and what type of approach that they will take to producing this product, if this can be known or inferred from their past actions.

3. The similarity or dissimilarity of this product or program to other recent experience, as to
   a. Relative design complexity
   b. Producibility
   c. Service life
   d. Maintainability
   e. Reliability
   f. Cost

If additional intelligence, or changes in the above, should become known during the estimating span, it should be shared with the quoters on some kind of systematic basis.

Certain other cautions to the front office in this general area of concern are treated in the next paragraphs.

TOP MANAGEMENT PARTICIPATION IN QUOTING

Determining Whether to Quote

Besides the part played by the high level people in making guidelines or intelligence data available to the quoters as the quote is underway, as mentioned earlier, they also should screen product ideas before formal quoting procedures are initiated to determine if the company really can benefit from producing the article. Questions impacting this decision include—

1. Is there a reasonable likelihood of turning a profit on the product?
2. Do we have the experience, people and facilities to do a good job on the product?
3. Will the production of the product interfere with other ongoing or proposed programs?
4. What is the likely effect on our reputation if we do a good or a poor job on the product?

Developing Intelligence for the Quote

Besides identifying similar programs, the activities of the competition, and the other factors mentioned earlier, management should also assist those quoting by providing, in readily usable form, data on past products, including:

1. Production volume
2. Costs (manhours, material costs, etc.)
3. Comparability (of Complexity & Function)
4. Schedules

My experience has been that the estimator will keep his own records of the above. However, the widespread use of computerized record systems could allow them to be produced without that kind of duplicative effort.

Screening the Quote

I am accustomed to calling this management review of the quote for internal consistency and overall cost compared to similar programs a "sell-off meeting." Actually, the screening for internal consistency of assumptions and estimates from various sources can be done beforehand by competent staff people, although their perspective will not be quite as broad as that of a top manager. It is of great importance that management not make across-the-board cuts in estimates on the sometimes mistaken assumption that everyone has padded his estimate by 10 or 20 percent. It is true that estimators do on occasion put a little more conservatism into their numbers than necessary, to allow a little room for negotiating, but in some instances where it is common knowledge that the competition is keen, they may have already trimmed their numbers to the bone. In general, management should discourage this practice of padding and cutting quotes so that the estimators will develop a stronger feeling of responsibility for the realism of the quote.

We have already discussed the usual procedure followed in planning for and developing a quotation, but several emphases need to be given to the actual estimating part of the process, which is the step that is most likely to go wrong.

PREPARATION OF THE ESTIMATE

First, as you estimate the requirements of your particular function in the plant, or of your particular component of the product, you will make assumptions that will affect the quotes for other functions or components. For example, if it is assumed that engineering has specified the location of certain holes at a tolerance which is more critical than that ordinarily expected, then the tooling organization will have to provide commensurately accurate means of locating the holes, thereby increasing the cost of producing them. It is extremely important that such assumptions be commonly known at the time that the individual quotes are being developed, prior to their submittal for coordination and pricing, because it usually is not practical, because of tight schedules, to send the individual quotes back to the originator to adjust for such reasons.

The estimator should be certain to develop the individual components of his estimate at the most extensive level of detail practical but this concern should not obscure the necessity for providing for such likely contingencies as waste, rework, and redesign. Having first developed the quote in this way the results of which we shall hereafter call the "Details Developed Quote," he should reconcile it in total to actual experience on similar programs. We shall call this second approach the "Experienced Based Quote." Then, if the newly calculated Details Developed Quote appears too low compared to past experience, the estimator should ask himself if he has omitted certain components or tasks from the quote, and if so, make the necessary adjustments. On the other hand, if the quote is too high, compared to past experience, he should either consider reducing it or else satisfy himself that the difference is justified in terms of new features in the design, increased cost of materials or components, or for some other valid reasons.

Finally, the estimator should very carefully document all of the assumptions for his quote, not omitting the assumed adherence to military or other specifications, the existence of data requirements including drawings and written and oral reports, the specification of product test requirements, or other factors that will have a significant effect on costs or that help describe exactly what is to be delivered to the customer.

We have been discussing the relative merits of building up our total details developed estimate from smaller estimates showing the most extensive level of detail, as compared to the totally experienced based quote using a more aggregative approach, where we simply quote approximately the same figure that we spent on a similar program. This latter approach would more likely be used for engineering type efforts, whereas the details developed quote would probably be used where parts manufacture or assembly is to be covered, and the totals would then be compared to past expenditures. The details developed quote has side benefits quite apart from the accuracy of the cost estimate.

MANAGEMENT BENEFITS FROM DETAILS DEVELOPED QUOTING

Details developed quoting better stimulates the necessary planning to adequately identify needed facilities for the new product. This is particularly important for specialized facilities with long lead times. At times, the company may want to reserve delivery for these in advance of an actual firm contract on the product by means of a letter of intent to the manufacturer.

It better allows the identification of specialized manpower needs, and stimulates detailed planning for recruitment, training, and securing of temporary help such as contract engineers, etc.

It better allows the assessment of current areas of interest of your customers, even if you happen not to be selected as the supplier on this product. This is not to say, of course, that you should bid on every available product just to gather information, but when you do decide to bid after careful consideration of your needs, you do secure this secondary benefit.

The quoter is encouraged to develop his estimates at the most extensive level of detail possible, at the same time paying careful attention to the overall size of the quote as compared with past experience. Furthermore, there are the resulting benefits just discussed of identifying special needs for facilities and manpower and of gathering intelligence on the customers' current interests. Several other benefits of getting the "big picture" can be named.

MANAGEMENT BENEFITS FROM ATTENTION TO THE OVERALL SIZE AND NATURE OF THE QUOTE

As the various parts of the quote are coordinated for internal consistency, better attention can be given to assessing the overall degree of complexity of the combined components of the product, particularly in the manner of how they interact. Symbolically, you shouldn't combine a Model T engine with a 1974 Cadillac body.

There can be a better understanding of the systems nature of the product, as brought out by the assumptions and estimates developed for the individual components and functions of the production effort. This, in turn, can lead to the benefits derived from:

1. Tradeoffs of design for cost, or of cost for design, or of one cost for a lesser cost as when additional computer time is substituted for engineering analysis time.

2. Tradeoffs of the increased time and expense of coordination activities for the scheduling benefits of additional subcontracting.

3. Establishment of better accountability, which is the determination of internally consistent goals for degree of design sophistication, level of costs, ease of manufacture, maintainability of the product, etc.

4. Establishment of a more logical system for tracking and controlling costs once the program is underway.

5. Better identification, during the progress of the program, of the problems of design, coordination, and cost, and of better information for making effective decisions to deal with these.

6. Better motivation of managers and individual contributors to the product resulting from the establishment of a system of accountability calling forth genuine contributions to company goals.

Estimating product engineering costs

by Dr. Phillip F. Ostwald • University of Colorado

The design of new products or redesign of existing ones often entails a heavy investment in engineering costs. This paper discusses methods for engineers to arrive at realistic assessments of product developmental costs so that management can better gauge the advisability of going ahead with a new product or redesigning an existing one.

On-going product effort is basic to a firm's survival, and its progress rests jointly upon research, engineering, manufacturing, marketing, and management. The hazard for new products is high since from some several hundred fresh ideas only one will be profitable. Likewise, many established products see their markets decline and, in some cases, disappear.

The traditional cost and price structure of an appliance is indicated by Fig. 1. Practices for estimating the elements other than engineering are well known. This paper will discuss four methods of estimating engineering costs. Such costs are critical to many new product designs and to a lesser degree to products undergoing redesign.

Engineering costs are defined as the total of all costs incurred in design to produce complete drawings and specifications or report. Included are the costs, salaries, and overhead for engineering administration, design costs, drafting, reproduction, engineering cost estimating, and purchasing and construction costs for prototype or developmental units.

Methods for estimating engineering costs vary depend-

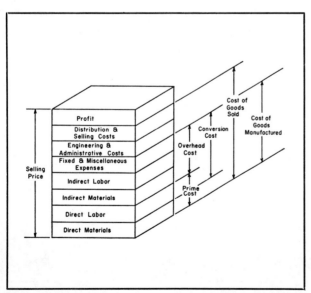

Fig. 1—In the traditional cost and price structure, estimating engineering costs is often most difficult.

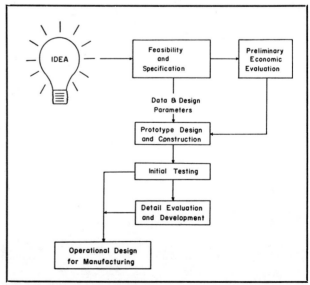

Fig. 2—A typical product development program requires careful economic evaluation right from the very start.

ing upon whether the product is a new appliance or existing one. Unfortunately, to distinguish precisely between what is a new product and that which is continuing is not always easy.

For instance, in the field of electric housewares and personal care products, development and marketing efforts deliver several new products to the marketplace annually. Yet, in many instances these products, although different in end use, are similar in design to existing ones and engineering may use the rather non-analytic comparison method or the line method described later in this paper to arrive at a highly accurate estimate of engineering cost.

On the other hand, a new model washing machine may look much like its predecessor and perform much the same function yet have entailed a large engineering cost because the product was redesigned "from the ground up." The new machine may feature a solid-state programmed timer, new drive mechanism, basket redesign and reselection of material, all representing a significant development investment which must be carefully estimated prior to project go ahead in order that the product ultimately be competitive in the marketplace.

Generally speaking, however, products which are in the continuing category offer fewer difficulties for manufacturing to estimate. Components are compared to the old design and classified as changed, added, or identical parts. Costs of the identical parts are found from records and may be altered to reflect future conditions. New estimates are prepared for the changed parts. Their composite estimate usually considers the cost of goods manufactured, or the bottom five layers of Fig. 1.

For new products, a typical product development program based upon the light bulb inspiration idea is dramatized by Fig. 2. Generally speaking again, these estimates are the hardest to figure for engineering and manufacturing. Overstatement of costs may result in an early death in the preliminary evaluation, while an understatement leads to unfortunate consequences later.

THE ESTIMATOR AND HIS TOOLS

Although engineering cost may be estimated by others outside of the engineering team, too much dependence is placed upon historical records which may have been crudely developed, and the estimate fails to reflect the spirit of the design problem. For this reason estimating done by the product design manager or an engineering cost specialist with past experience in engineering is prefered. This specialist provides management information to help make product development decisions, while accounting provides information to assist in evaluating decisions made earlier.

In order to do his job well, the engineering cost specialist should have at his disposal various kinds of information including budget and variance reports, man-day summaries, policy costs, and manufacturing standards. For most situations, the basic document is the budget.

Expense Account	Product Engineering	Research & Development	Engineering Services
Salaries	$50,000	$25,000	$10,000
Total salaries	$50,000	$25,000	$10,000
FICA	2,700	1,350	540
Workmen's compensation	500	250	100
Group insurance	1,000	500	200
Pension fund	1,250	625	250
Fringe benefits	1,000	500	200
Taxes	1,000	500	200
Total payroll related	$ 7,450	$ 3,725	$ 1,490
Overtime	750	600	400
Office supplies, reproductions	3,000	1,000	500
Spoiled materials	150	150	100
Freight charges	375	450	300
Traveling	750	750	500
Books	75	75	50
Memberships	75	75	50
Telephone	750	750	500
Postage	150	150	100
Professional services	1,050	600	400
General expenses	750	1,200	800
Depreciation assets	750	750	500
Rent furniture	375	600	400
Building space charges	1,125	900	600
Total expenses	$10,125	$ 8,050	$ 5,200
Engineering errors	6,000	20,000	
Manufacturing costs	1,500	30,000	
Total shop costs	$ 7,500	$50,000	0
Total expected costs	$75,075	$86,775	$16,690

Table I—The expense budget itemizes expenditures that are expected to be incurred during the budget period.

Product Identification	Variance to Date	Cost to Date	QUARTERS 1 Eng	1 Mfg	2 Eng	2 Mfg
444	+ 486	3,486	5,000	10,000		
3806	+ 297	9,297	3,000	5,000	8,000	5,000
443	- 375	4,625	3,000	5,000	5,000	5,000
2211	+ 22	1,322	11,000	25,000		
7582					5,000	
7543					5,000	
7542						
7547	+ 324	824	3,000	5,000	5,000	10,000

Table II—A product budget will project expenses and tabulate current costs for each product being studied.

	1st Qtr.	2nd Qtr.	3rd Qtr.	4th Qtr.	Year
Product Engineering					
Salaries	$50,000	$47,000	$47,000	$50,860	
Payroll related expenses	7,450	4,700	4,700	5,080	
Expenses	10,125	1,000	22,650	8,150	
Errors in manufacturing	7,500	10,000	7,750	7,900	
	$75,075	$62,700	$82,100	$71,930	$291,805
Research & Development					
Salaries	$25,000	$28,000	$29,000	$25,000	
Payroll related expenses	3,725	2,800	2,900	2,500	
Expenses	8,050	8,000	7,850	22,350	
Shop costs	50,000	20,000	25,000	30,000	
	$86,775	$58,800	$64,750	$79,850	$290,175
Engineering Services					
Salaries	$10,000	$10,000	$13,400	$13,400	
Payroll related expenses	1,490	1,000	1,340	1,340	
Expenses	5,200	5,000	6,860	6,860	
	$16,690	$16,000	$21,600	$21,600	$ 75,890
Total for fiscal year					$657,870

Table III—The engineering department's summary budget combines individual factors into an overall picture.

Budget Information Needed

The budget is a written plan covering the activities for a definite future time expressed in dollar units. Codification of various cost categories for the budget are important so as to reduce ambiguous charging; whereas, budget definitions establish engineering cost centers and accounts to allow the placing and tracing of costs.

A simplified engineering budgeting scheme has three types—expense, product, and summary. The expense budget is a schedule of expenses which the department plans to spend during the budget period, a quarter in the case of Table I. Complexity of the expense budget depends upon the amount of information available and the degree of control desired. There are four main sections in this example: salaries, payroll overhead expenses, supplies, and manufacturing expenses.

The product budget, Table II, lists current engineering products and estimates for several future quarters. The projected expenses and the sum total cost to date are shown for each product. Total variance for each product can be a part of the product budget so that comparisons may be made between the budget target and actual cost. A summary engineering budget, Table III, is the master document, but it is not associated with specific products.

Procedures should be organized to supply actual expenditures to the managers who are responsible for the departmental budgets. The budget can serve in a dual capacity since reported variances, or departures from budgeted amounts, can be used by the cost engineering specialist to forecast man-hours and dollars for future product design. Ideally, this information should periodically come back in the same form as the budget was submitted — broken down by cost centers and expense accounts for the expense budgets and by product number for the product budgets. Significant differences between actual and budgeted costs can indicate either poor control of the department or a poor attempt at budgeting or both.

Man-Day Data

Man-day summaries are another form of information available to the cost engineering specialist. These are data that relate design job descriptions to man-days. Examples could be "3 Drawings Size-A per day" and "Design small overriding clutch per 21 days" (whether one or more men). Notwithstanding the abuses that this information can cause, the enginering cost specialist needs it for clearer understanding and better estimates.

The difficulties in obtaining such information are due to the fact that typical professional engineering design jobs lack adequate tasks definition, and procedures for measuring these tasks are not defined. Where job recording systems have been used for some time, the engineering cost specialist has developed a "personal" understanding of what various product designs entail in terms of man-days. While this understanding may not be always accurate, it aids the estimator in getting a feel for a particular product or component design.

One of the better ways to determine man-day standards has the manager and design engineer jointly contribute information. They indicate the job content, days or hours worked, and a description of the person doing

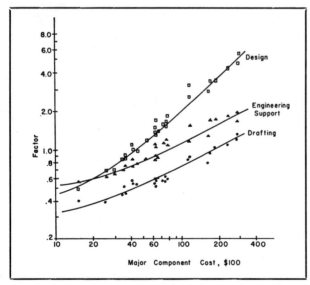

Fig. 3—Typical "Factor Method" chart relates engineering cost to a parameter such as major component cost.

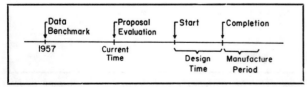

Fig. 4—Time scale adjustments are necessary to translate bids into actual costs at the time of production.

Type of Labor	Task hrs.	Rate hr.	Extended Design Cost
ENGINEER, SENIOR DESIGN			
ENGINEER, DESIGN			
SCIENTIST, RESEARCH			
TECHNICIAN, ELECTRICAL			
DESIGNER/ENG. AIDE			
TECH WRITER			
ILLUSTRATOR			
DRAFTSMAN			
PROVISIONING SPECIALIST			
MODEL SHOP LABOR			
PROGRAMMER AIDE			
DESIGN MATERIALS			
OUTSIDE ENGINEERING SERVICES			
TOTAL DESIGN ENGINEERING LABOR			

DESCRIPTION _____
CUSTOMER _____ INQUIRY OR QUOTE NO. ____ DATE ____
BASED ON QUANTITY OF _____ DURING PERIOD OF _____

Table IV—Product engineering estimating forms such as this serve for products that follow established patterns.

the job. These standards have greater value if the time to perform the task is relatively constant and unaffected by changes. While this time-value consistency is unlikely, the record includes backup data to assist the engineering cost specialist in determining deviations for future tasks.

COST ESTIMATING METHODS

Armed with budget information and man-day standards the cost engineering specialist or project manager has four methods of estimating engineering costs at his disposal. These methods — conference, comparison, line item, and factor — become progressively more numeric and depend less upon intuitive judgment; however, each one has certain merits.

- *Conference Method.* The conference method is a nonquantitative method of estimating which provides a single value of anticipated cost using the experience of the estimator(s). The method relies upon judgment, individual or collective, of contrasting differences between previously determined designs and their associated costs to an unknown but to be determined estimate with its future design.

Various gimmicks can be used to sharpen judgment. A "hidden card" technique has members of the estimating team reveal personal value. This could provide a consensus. If agreement is not initially reached, discussion and persuasion are permitted as influencing factors. The major drawbacks of the conference method are the lack of analysis and a trail of verifiable facts leading from the estimate to the governing situation.

- *Comparison Method.* The comparison method is similar to the conference one with the addition of a formal logic. An unsolvable or exceptionally difficult design and estimating statement is designated problem A, and a simpler design problem for which an estimate is known is constructed. The simple problem is called problem B. The simple problem might arise from a manipulation of the original design or relaxation of the technical constraints upon the original problem.

One gains information by branching to B as various facts exist about B. Indeed, estimate B may need only a minor restructure of data to allow comparison; however, it must be selected to bound the original problem A in the following way:

$$C_A (D_A) \leq C_B (D_B)$$

where $C_{A,B}$ = value of the estimate for design A and B; and $D_{A,B}$ = design A or B and D_B must approach D_A as nearly as possible.

The sense of the inequality in the expression is for a conservative stance. It may be management policy to estimate or price slightly higher at first, and, after the detail estimate is complete with D_A fully explored, the cost for the unknown problem A is generally less than the original comparison estimate. An additional lower bound is also possible, which assumes a similar circumstance for a known design, C. The unknown design, A, is bracketed by C_B and C_C.

While there is no clear boundary that distinguishes the various characteristics of estimating methods one thing is certain: Confidence in these two methods is weak, and invariably these practices can be improved by the following analysis methods which become progressively more quantitative.

- *Line Item Estimate.* When products are like the existing products or when there is reasonable assurance about the state of the art in terms of engineering, the line item estimating technique may be adopted. For these conventional situations, engineering is estimated in hours using similar and past budgets and variance reports.

A recap sheet like Table IV is typical of this approach. The design engineering estimate is categorized by type of labor, task hours, rate per hour, and finally summarized in total design and engineering labor. The total of this cost is increased by ratios reflecting overhead cost for the engineering department. Rather than a specific hourly estimate for an engineering labor grade, a lump sum estimate for engineering department cost centers may be determined. Total engineering cost is arrived at by summing and subsequently ammortized to the product.

Year	Fabrication and Assembly Manufacture Index	Drafting Index	Engineering Support Index	Design Index
1950	74.0	83.2	75.6	81.7
1957	97.5	98.5	94.2	94.6
1958	99.8	99.1	100.1	100.5
1959	101.3	101.3	99.2	100.0
1960	102.0	104.2	101.3	100.0
:	:	:	:	:
1969	111.8	116.—	125.7	116.2
1970	114.3	119.7	131.2	119.7
1971	115.4	124.3	136.2	124.3
1972*	118.0	127.5	141.0	127.0
1973*	119.0	131.0	146.0	128.0
1974*	121.5	133.0	149.0	130.5

*Extrapolated value

Engineering Category	1957-1959 Factor	1957-1959 Cost[1]	1974 Index	1974 Estimated Engineering Costs
1. Drafting	0.9	$14,490	133.0	$19,271
2. Engineering support	1.5	24,150	149.0	35,984
3. Design	3.1	49,910	130.5	65,133
4. Other costs Programming Technical Writing				
			Total	$128,388

[1] Major component cost ≟ $16,100

Table VI—Using the Factor Method, a table such as this can be drawn up for cost estimating.

Table V—Indexes such as these are used to find the reference year cost (1957-1959 = 100).

- *The Factor Method.* For some designs there is a natural dependence of engineering cost on some engineering parameter such as component cost, or current load for a switch or horsepower (torque) for an electric motor. Extensive background studies where a multiplying factor is statistically correlated to a major component parameter is necesary.

Fig. 3 is an actual correlation of cost to a major component for a non-appliance design where component costs happen to be significantly higher than for an appliance design. This choice of example, however, does not negate the usefulness of the factor method for estimating engineering costs for the redesign of an existing appliance. If component cost proved to be correlated to engineering costs for an appliance component, the component cost (abscissa, Fig. 3) would be deflated and the weighting factor (ordinate) inflated, and the functional relationship might be altogether different from that of Fig. 3.

The factor method finds engineering cost by using the equation:

$$C = \sum_{i=1}^{n} f_i D_i,$$

where
- C = cost of engineering design
- D_i = design segment i (design, engineering support, etc.)
- f_i = weighting factor relating engineering cost to component parameter.

The general configuration of the product or a flow chart or specification sheet are the usual input data for the factor method. From the flow sheet the basic items of the product are identified but not designed. For instance, the motor horsepower of an appliance may be selected. Its fabricated and assembled cost is estimated by manufacturing or from quotations by vendors.

Costs for drafting, engineering support, and product design are then determined based on the use of a graph such as Fig. 3. Costs which are not component parameter sensitive such as programing or technical writing must be estimated separately by other methods.

Factor data are collected for some past period, with 1957-1959 frequently chosen, as this is the period that serves as a bench mark for many government indexes. Time-scale adjustments (see Fig. 4) follow this course. First, the basic item is submitted for vendor bid or estimated internally at today's prices. With one or more values for the basic item, the average value is deflated to the reference point like 1957 for finding the factors. Using this adjusted cost, the curve is entered and factors are found. It is then indexed or inflated forward to the time of anticipated design.

A simplified example illustrates the factor method. Assume that an existing product is to be redesigned. From preliminary designs and specification sheets a gear train is specified, and correlating analysis has shown consistent dependence of this component's cost to engineering cost. Consequently, the factor method is applicable.

External bids and an internal estimate were received for which the rough average of the component cost is $19,000. Inasmuch as the bid is current time (1972), the estimate for the major component is deflated using cost

Dr. Phillip F. Ostwald *was graduated from the University of Nebraska in 1954 and received his PhD from Oklahoma State University in 1966. He joined the University of Colorado in 1966 in the department of Engineering Design and Economic Evaluation and teaches courses such as Operations Research, Operations Optimization, Advanced Engineering Economics, and Design Estimating. In addition, he has presented numerous short courses for industry, societies, and the government. Dr. Ostwald has over 30 publications, including a forthcoming textbook titled* Cost Estimating Systems, Projects, Products, and Operations.

indexes such as those found in Table V to find the reference year cost (C_{1957}):

$$C_{1957} = 19{,}000 \left(\frac{100.0}{118.0}\right) = 16{,}100$$

With this value, factors are determined for the three departmental categories using Fig. 3 and are entered as 1957-1959 Factors in Table VI. Extensions for drafting, engineering support, and design are indicated next. These values are then inflated to the point where design time is anticipated to start, say 1974. For example, drafting cost for 1974 is determined as

$$C_{1974} = 14{,}490 \left(\frac{133.0}{100.0}\right) = 19{,}271.$$

Certain costs may be independent of the major component and are estimated separately. For this example an approximate sum of $128,500 is the final estimate.

It should be emphasized that for this illustrative example component cost was found to be a useful engineering parameter, but that for many components other parameters would prove to be more useful. For instance, if the component is an electric motor, engineering costs might be less sensitive to final motor cost than to required torque, motor size or allowable operating temperature or some other motor parameter.

In conclusion, conference, comparison, line item and factor methods are useful for estimating engineering costs to aid management in enhancing the product line picture or to add a dollar amount to the product cost to recover engineering costs. For this cost-evaluation effort, it is likely that the estimating specialist will be an engineer knowing the problems of design and that he can choose among these methods based on the nature of the product design and the availability of supportive material. — AE

Adapted for APPLIANCE ENGINEER *from a paper presented at the Design Engineering Conference, May 8-11, 1972, and published by special permission of the author and the Design Engineering Div. of the American Society of Mechanical Engineers.*

CHAPTER 4

LEARNING CURVE

Learning Curve

9.4 LEARNING CURVE

It is frequently recognized that repetition with the same operation results in less time or effort expended on that operation. In fact this improvement can be sufficiently regular to become predictive through ordinary estimating techniques. The observed characteristic of the improved performance is called *learning*. The first applications of learning were in airframe manufacture, which found that the number of man-hours spent in building a plane declined at a constant rate over a wide range of production. Figure 9.4 describes this phenomenon.

Other names abound for the learning curve including terms such as the manufacturing progress function and the experience or dynamic curve. They suggest that cost can be lowered with increasing quantity of production or experience. Knowing how much product cost can be lowered and at what point

FIGURE 9.4

LOG-LOG PLOT OF INDUSTRY AVERAGE UNIT CURVE FOR CENTURY SERIES AIRCRAFT. (From Aeronautical Material Planning Report Data)

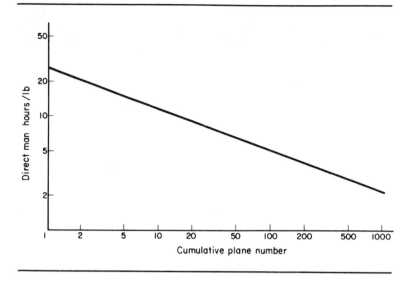

learning is applied in the estimating procedures are the reasons for studying learning prior to product estimating methods.

Applications of the learning curve are found in procurement, production, and the financial aspects of a manufacturing enterprise. In purchasing, a function can be used to negotiate purchase price or it may be used for the *make or buy* decision. Learning has been applied to equipment loading and manpower schedules. In cost estimating, decisions related to bidding, pricing, and capital requirements are based in part on the concept. Contract negotiation is sometimes re-opened after a first satisfactory model has been produced. With the experience of time and cost for the prototype unit known, the contract for later models is based on learning reductions. Aerospace firms do this with the Air Force and precision of $\pm 3\%$ has been reported.

There are explanations for this behavior and verification has been uncovered by independent researchers and companies. Principally, the reduction is due to direct-labor learning and the management process. The direct-labor learning process assumes that as a worker continues to produce, it is natural that he should require less time per unit with increasing production.

The management processes are those engineering programs which improve production, encourage quality, reduce design complexity, create technology progress, and foster product improvement. In short, they are those management programs that are considered extraordinary and exceptional to a particular product. These management programs inspire time and cost reduction and are credited with time and cost reduction of a product additionally beyond that which direct-labor learning would provide.

Some experience suggests that the operator is responsible for approximately 15% of the total reduction, while management and their programs contribute the remaining 85%. In the manufacturing industries the 85% has been broken down into 50% due to the product engineering endeavors, while manufacturing and industrial engineering activities are credited with the remaining 35%.

Ships, aircraft, computers, machine tools, and apartment and refinery construction have in common high cost, low volume, and discrete item production and can be treated by learning. Although the same principle applies to TV production, for instance, the effects may take years to uncover because of the large production volume. The learning curve is usually not applied to high-volume or low cost products.

9.4-1
Learning Curve Theory[1]

The learning curve model rests on the following observations:

1. The amount of time and its cost required to complete a unit of product are less each time the task is undertaken.
2. The unit time will decrease at a decreasing rate.
3. The reduction in unit time follows a specific estimating model such as the negative exponential function.

To state the underlying hypothesis, the direct labor man-hours necessary to complete a unit of product will decrease by a constant percentage each time the production quantity is doubled. While the hypothesis stresses only time, practice has extended the concept to other types of measures. A frequent stated rate of improvement is 20% between doubled quantities. This establishes an 80% learning curve and means that the man-hours to build the second unit will be the product of 0.80 times that required for the first. The fourth unit (doubling 2) will require 0.80 times the man-hours for the second; the eighth unit (doubling 4) will require 0.80 times the fourth; and so forth. The rate of improvement (20% in this case) is constant with regard to doubled production quantities, but the absolute reduction between amounts is less. This is the reason why *follow-on* costs in low production are noticeably lower than original costs. The learning curve may be defined if the number of direct labor hours required to complete the first unit is established and if the subsequent rate of improvement is specified. Alternatively, the learning curve may be defined if direct-labor man-hours for a downstream unit and the learning curve rate are estimated. Other possibilities for defining the curve can be selected. The number of direct-labor hours required to complete the first production unit depends on these circumstances:

1. The previous experience of the company with the product. If it had little or no experience, the first unit time would be greater than for a product with considerable experience.
2. The amount of engineering, training, and general preparations that the organization expends in preparation for the product. In some cases the first several units are custom-made and tooling is not made until more sales can be assured. This "hard-way" production would inflate the first unit time.
3. The characteristics of the first unit. Large complex products would be expected to consume more direct-cost resources than something less complex.

The notion of constant reduction of time or effort between doubled quantities can be defined by a unit formula:

$$T_N = KN^s \tag{9-4}$$

where T_N = effort per unit of production, such as man-hours or dollars required to produce the Nth unit
N = unit number

[1]Excerpts from W. J. Fabrycky, P. M. Ghare, and P. E. Torgersen, *Industrial Operations Research*, Prentice-Hall, Inc., Englewood Cliffs, N.J., 1972, pp. 172–200.

$K =$ constant, or estimate, for unit 1 in units compatible to T_N
$s =$ slope parameter or a function of the improvement rate

The slope parameter is negative because the effort decreases with increasing production. Figure 9.5 is the linear unit progress curve for 80% learning with cartesian coordinates, while Figure 9.6 represents a log-log curve.

FIGURE 9.5

80% MANUFACTURING PROGRESS FUNCTION WITH UNIT NUMBER AT 100 DIRECT-LABOR MAN-HOURS

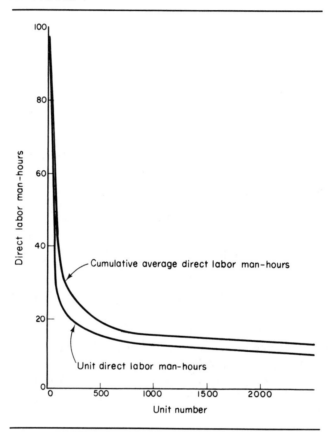

To understand the graphic presentation of the learning curve on logarithmic graph paper, first compare the characteristics of arithmetic and logarithmic graph paper. On arithmetic graph paper equal numerical differences are represented by equal distances. For example, the linear distance between 1 and 3 will be the same as from 8 to 10. On logarithmic graph paper the linear distance between any two quantities is dependent on the ratio of those two quantities. Two pairs of quantities having the same ratio will be equally spaced along the same axis. For example, the distance from 2 to 4 will be the same as from 30 to 60 or from 1000 to 2000.

The learning curve is usually plotted on double logarithmic paper, meaning that both the abscissa and the ordinate will be a logarithmic scale. For an exponential function $T_N = KN^s$ (the general class of learning curve functions)

the plot will result in a straight line on log-log paper. Because of this, the function can be plotted from either two points or one point and the slope, e.g., unit number 1 and the percentage improvement. Also, by using log-log paper the values for a large quantity of units can be presented on one graph, and these can be read relatively easily from the graph. Arithmetic graph paper, on the other hand, requires many values to sketch in the function.

FIGURE 9.6

LINEAR UNIT PROGRESS CURVE WHERE
ϕ = 80% AND UNIT 1 = 100 HOURS

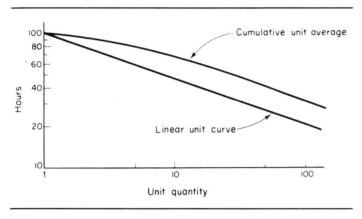

The slope constant s is negative because the time per unit diminishes with production. The units for T_N can be expressed in any of several compatible dimensions. Time or cost is frequently used, but when cost is used the underlying basis should be estimated with care.

The cartesian-coordinate curve representation indicates the range of major reduction. It is evident that any special management program for Figure 9.5 may be unnecessary beyond the five-hundredth unit or so, which is approximately where the slope begins to level out. This judgment is an evaluation of input resources to output success and points up that the savings occur early in the life of some products. As the product becomes older it becomes less susceptible to improvements.

The exponent s is defined (recalling the double quantity concept) as

$$s = \frac{\log \phi}{\log 2} \tag{9-5}$$

where ϕ = the slope parameter of the learning curve function.

Application of Equations (9-4) and (9-5) can be illustrated by an 80% learning curve with unit 1 at 1800 direct-labor man-hours. Solving for T_8, the number of direct-labor man-hours required to build the eighth unit gives

$$T_8 = 1800(8)^{\log 0.8 / \log 2}$$
$$= 1800(8)^{-0.322} = \frac{1800}{1.9535} = 921 \text{ hours}$$

A table of decimal learning ratios to slope constant is given as

Slope Parameter, ϕ	Exponent, s
1.0 (no learning)	0
0.95	−0.074
0.90	−0.152
0.85	−0.234
0.80	−0.322
0.75	−0.415
0.70	−0.515
0.65	−0.621
0.60	−0.737
0.55	−0.861
0.50	−1.000

For unit 1 requiring 1500 hours and a projected learning of 75%, the time for unit 90 is

$$T_{90} = 1500(90)^{-0.415} = \frac{1500}{6.471} = 232 \text{ hours}$$

Tables shorten the calculations. Appendixes III and IV are unit and cumulative values, respectively.

9.4-2 Finding the Learning Curve from Two Points

It may be desired to find out if learning has materialized, and if it has, to determine the function for which other unit estimates can be found. Where only two points are specified it may be desirable to find the learning curve that extends through them. Let the two points be specified as (N_i, T_i) and (N_j, T_j). At each point

$$T_i = KN_i^s \quad \text{and} \quad T_j = KN_j^s$$

Dividing the second into the first gives

$$\frac{T_i}{T_j} = \left(\frac{N_i}{N_j}\right)^s \tag{9-6}$$

and taking the log of both sides we have

$$\log \frac{T_i}{T_j} = s \log \frac{N_i}{N_j}$$

$$s = \frac{\log T_i - \log T_j}{\log N_i - \log N_j} \tag{9-7}$$

K may be found by substituting s into $T_i = KN_i^s$ and solving for K:

$$\log T_i = \log K - s \log N_i \tag{9-8}$$

This has a linear form like $y = a + bx$, one of the linear models described in Chapter 5. It is seen that on log-log paper the intercept is K while the slope of the line is equal to $-s$.

An example will now illustrate this concept. A company audited two production units, the twentieth and fortieth, and found that about 700 hours and 635 hours were used, respectively. Now at the seventy-ninth unit, they want to estimate the time for the eightieth unit:

$$s = \frac{\log 700 - \log 635}{\log 20 - \log 40}$$

$$= \frac{2.8451 - 2.8028}{1.3010 - 1.6021} = \frac{0.0423}{-0.3011} = -0.1405$$

and

$$\log \phi = s \log 2$$
$$= (-0.1405)(0.3010) = 9.9577 - 10$$

Taking the antilog of both sides gives $\phi = 0.907$. The percentage learning ratio is 90.7%. Using the data for the twentieth unit,

$$700 = K(20)^{-0.1405}$$
$$\log 700 = \log K - 0.1405 \log 20$$
$$\log K = 2.8451 + 0.1405(1.3010) = 3.0279$$

Taking the antilog, $K = 1066$ hours. The learning curve function is

$$T_N = 1066 N^{-0.1405}$$

The unit time for any unit can now be calculated directly. For the eightieth unit

$$T_{80} = 1066(80)^{-0.1405} = 576 \text{ hours}$$

This can be confirmed by using Appendix III and interpolating. It should be observed that

$$\frac{T_{40}}{T_{20}} = \frac{T_{80}}{T_{40}} = 0.907$$

9.4-3 Cumulative Curve

The unit formulation can be extended to other types of functional models. It may be necessary to determine a cumulative average number of direct-labor man-hours. This can be found by the cumulative total and is

$$T_T = T_1 + T_2 + \cdots + T_n = \sum_{i=1}^{n} T_i \qquad (9\text{-}9)$$

and its average is

$$\sum_{i=1}^{n} \frac{T_i}{n} \qquad (9\text{-}10)$$

A good approximation of the cumulative average number of direct-labor man-hours for 20 or more units is given by

$$V_n \doteq \frac{1}{(1+s)} K N^s \qquad (9\text{-}11)$$

where $V_n \doteq$ cumulative average of number of direct-labor man-hours.

The cumulative average curve is above the linear unit curve. See Figure 9.6.

Unit 1 is estimated to require 10,000 hours and production is assumed to have an 80% learning curve. What will be the unit direct-labor man-hours, the cumulative direct-labor man-hours, and the cumulative average direct-labor man-hours for unit 4?

$$T_1 = 10,000$$
$$T_2 = 10,000(2)^{-0.322} = 8000$$

$$T_3 = 10{,}000(3)^{-0.322} = 7021$$
$$T_4 = 10{,}000(4)^{-0.322} = 6400$$

The cumulative direct-labor man-hours are

$$\sum_{i=1}^{n=4} T_i = 31{,}421$$

while the average using Equation (9-10) is

$$V_n = \frac{1}{4}(31{,}421) = 7855$$
$$= \frac{6400}{(1 - 0.322)} = 9440$$

Formula (9-11) is not accurate in this quantity range since it is on the upper left-hand hump of the curve. This can be seen by examining Figure 9.6.

9.4-4 Least-Squares Fit

No estimating device despite its attractiveness can be used indiscriminately. For this reason it is desireable that management insist on occasional audit exercises that measure actual learning as a function of production. By using the results from a least-squares fit study, estimating is able to compare actual learning against predictions. These actual plots have a two fold purpose. They tend to prevent abuses and encourage additional effort as a need is shown, or they reduce effort as various programs are brought under control.

One means of past data evaluation is to plot the data on log-log graph paper and sketch a straight line through the data. However, it is better to adopt a least-squares fit as first described in Chapter 5. The general learning curve can be expressed as a logarithmic straight line of the form $\log T_n = \log K + s \log N$. The method of least squares will yield

$$s = \frac{M \sum \log N \log T - \sum \log N \sum \log T}{M \sum (\log N)^2 - (\sum \log N)^2} \tag{9-12}$$

$$\log K = \frac{\sum \log T \sum (\log N)^2 - \sum \log N \sum \log N \log T}{M \sum (\log N)^2 - (\sum \log N)^2} \tag{9-13}$$

where M = sample number. An example illustrates the procedure. Five data points have been collected as

Unit, N	Time, T
10	510
30	210
100	190
150	125
300	71

The method is described by Table 9.3. For the data the learning slope is 70% and the initial derived first unit time is 1556 hours. An estimating model is then constructed as

TABLE 9.3

LEAST SQUARES ANALYSIS OF ACTUAL COST DATA
TO FIND INITIAL VALUE AND LEARNING SLOPE

Given $T = KN^s$
or $\log T = \log K + s \log N$ which is of the form $y = a + bx$
where $y = \log T$, $a = \log K$ intercept, $b = s$ slope, $x = \log N$, and M = sample size

Unit, N	Man-Hours, T	$x = \log N$	$y = \log T$	$(\log N)^2$	$\log N \log T$
10	510	1.0000	2.7076	1.0000	2.7076
30	210	1.4771	2.3222	2.1818	3.4301
100	190	2.0000	2.2788	4.0000	4.5576
150	125	2.1761	2.0969	4.7354	4.5631
300	71	2.4771	1.8513	6.1360	4.5859
		9.1303	11.2568	18.0532	19.8443

$$s = \frac{M \sum \log N \log T - \sum \log N \sum \log T}{M \sum (\log N)^2 - (\sum \log N)^2}$$

$$= \frac{5(19.8443) - (9.1303)(11.2568)}{5(18.0532) - (9.1303)^2} = -0.515,$$

$$\log K = \frac{\sum \log T \sum (\log N)^2 - \sum \log N \sum \log N \log T}{M \sum (\log N)^2 - (\sum \log N)^2}$$

$$= \frac{11.2568(18.0532) - (9.1303)(19.8443)}{5(18.0532) - (9.1303)^2} = 3.19207,$$

antilog $K = 1556$

$T = 1556 N^{-0.515}$

$\log \phi = s \log 2 = -0.515(0.30102) = -0.15503$

antilog $\phi = 0.6697$ or $\phi = 69.97\%$

$$T_N = 1556 \, N^{-0.515}$$

This is a customary analysis used to determine past learning performance. Too often estimators are prone to use learning slopes without an adequate test of the firm's experience.

9.4-5 Specific Cost Applications

In predicting costs it is useful for estimators to know the recurring costs of production and the learning slopes for the several aspects of production, e.g., manufacturing direct labor, raw material, manufacturing engineering, tooling, quality control, and other indirect charges. In our next application we consider learning for direct labor and factory overhead. Materials are now assumed to be insensitive to the learning curve.

$$C_i = \frac{KN^s}{s+1}(C_{dl}) + \frac{KN^s}{s+1}(C_{dl})(OH) + C_{rm}$$

$$= \frac{KN^s}{s+1}(C_{dl})(1 + OH) + C_{rm} \qquad (9\text{-}14)$$

where C_i = product cost per unit for product i
C_{dl} = direct-labor hourly rate
OH = overhead rate including engineering, tooling, quality control, and other indirect charges expressed as a decimal of the direct-labor hourly rate

Our example considers a situation in which 200 units are to be produced. $C_{dl} = \$4$ per hour, raw material cost is $250, and the overhead rate is 50%. The first unit is estimated to require 350 direct-labor man-hours, and a 90% learning curve slope is thought applicable. The average cost per unit is

$$C_i = \frac{350(200)^{-0.152}}{0.848}(4)(1 + 0.50) + 250 = \$1357$$

For a lot of 200, the total cost is $200(1357) = \$271,359$.

Now we examine a problem having several diverse items that can be estimated by a learning curve approach. Some factors can be associated as recurring and identified to production quantity. Portions of these same items are also insensitive to quantity. Engineering can be isolated in this fashion. The first unit will require the bulk of design, but subsequent units, if a customer and or specifications necessitate, also require additional design on a per unit basis such as *engineering-change orders*. The latter portion can be subjected to a learning curve philosophy.

Table 9.4 is an illustration of a design for 150 units. Selected cost items have been broken out and divided into the categories of fixed and variable. The learning slope is estimated on the basis of a similar past performance. Extensions to a total line cost are straightforward. Other major cost categories which were not considered via learning curves are then added as a percentage of total estimated costs. Finally a total price is determined for the 150 units.

Learning Curves in Manufacturing Cost Estimates

By Raymond B. Jordan
Administrator—Cost Systems and Studies
General Electric Company

During the past ten years the use of the learning or progress curve has taken much of the drudgery and frsutration from the field of cost estimating. Long-range cost estimates have always suffered from the lack of a tool by which the cost progress to be experienced during the production period could be predicted empirically rather than resorting to intuition, flying-by-the-seat-of-the-pants, or some other unscientific method.

Since 1936, when T. P. Wright published his generic article,"Factors Affecting the Costs of Airplanes," in the Journal of Aeronautical Sciences, the learning curve technique has been available to the airframe industry but it was not until the era prior to 1970 that its use expanded significantly to other industries.

During the decade of the 1950's a technique known as formula estimating was given wide publicity. This method involved separation of past historical data into fixed and variable relations through the least squares procedure to fit the normal equation of the straight line. The fallacy involved in this method was the basic assumption that costs follow a straight line, an assumption already refuted by the T. P. Wright article. Experience with the airframe industry during World War II had already proved that costs are hyperbolic in shape rather than following a straight line.

The old formula estimating technique does have its place in estimating, however. It is useful, for example, in estimating the cost of end items such as mass-produced fractional horsepower motors which have reached the end of the learning curve. A least squares regression study will determine the cost differential between various sizes of motors in which the formula derived is similar to the following:

Cost = Base Cost (Fixed) + Variable (Number of Coil Windings)

The use of the above method has been employed with success for quick estimating by companies which mass produce a large volume of end items in which there is a demand for many tailor-made items in which the difference from one model to another is not great.

The purpose of this article is to explore the use of the learning curve technique in long-range cost estimates and call attention to some of the problems and pitfalls involved therein.

GENERAL USE OF THE TECHNIQUE

Negotiations between contractor and customer are expedited when learning curves are used because learning curve calculations serve as a common denominator from which to begin negotiation discussions. In business involving contracts with the Department of Defense or other government agencies, the use of the learning curve is recommended because government negotiators are using the technique to evaluate cost proposals. The Defense Contract Audit Agency (DCAA) is training employees in the fundamentals of the technique at their Memphis, Tennessee, educational facility in order to assist the negotiators. The DCAA has the added advantage of using time-sharing computer facilities to determine the slope of the proposal and quickly perform extrapolations.

THE LEARNING CURVE THEORY

Briefly, the learning curve theory involves the principle that repetition of an operation decreases the time involved in performing the operation. The thoery holds that every time the production of an item is doubled the cost or time to perform the operation decreases by a fixed percentage representing the degree of learning. For example, a 90 per cent curve means that the 200th unit should take only 90 per cent of the time taken by the 100th unit.

There are two basic theories used with each theory having its large group of followers. One theory holds that on log-log paper, the cumulative average line is straight and the individual unit line curves from unit one until unit ten from which point it parallels the cumulative average line. The other theory believes that the individual unit line is straight and the cumulative average curves. The merits of one theory versus the other will be treated separately within this paper. It is important, however, that the user identify the theory used so that there will be no misunderstanding during negotiations.

BASIS FOR ESTIMATING

The method used by many cost estimators today begins with determining the expected cost for an individual unit or a block at some period of time early in the production period. The unit or block selected can be the actual experienced cost or the estimator may forecast the cost which will be realized at a point when unusual starting costs have been eliminated and tooling problems have been resolved. For purposes of this article, the 250th unit has been selected.

Once the unit and its cost have been selected, the projections may be made on log-log paper using prior learning rates based upon experience gained from similar products. Figure One illustrates a hypothetical situation in which a cost of $100 at the 250th unit has been projected on a 90% curve. The theory that the cumulative average is straight has been used in the illustration. A feature of the learning theory holds that there is a constant relationship between the cumulative average line and the individual unit line. The mechanics of drawing the curve are as follows:

CONSTRUCTING THE CURVE

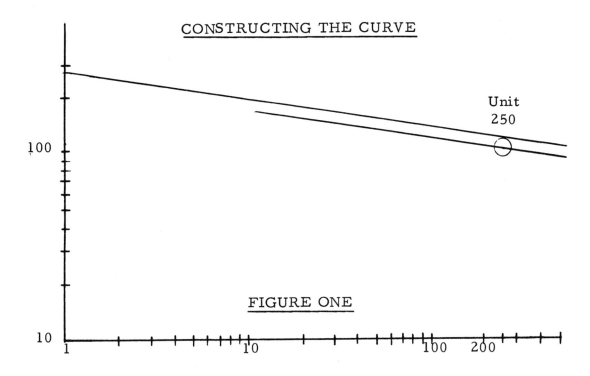

FIGURE ONE

1. Locate the 250th unit at a value of $100.
2. Locate the cost of the 500th unit at a value of $90.
3. Locate the cost of the 125th unit at $111, arrived at by dividing $100 by 0.9. Normally, two points should suffice by which to draw the line. It is recommended that three points be used in order to prevent inaccuracies or errors in plotting which will be revealed if the three points do not lie along a straight line.
4. Extend the unit line to the left as far as unit 10 and as far to the right as required.
5. Divide each of the three point values by 0.848, the factor for converting between the individual unit line and the cumulative average line.
6. Draw the cumulative average through the points plotted from step 5, extending the line to the left until it intercepts the vertical axis at the theoretical value for unit one.

USING THE LOG-LOG GRAPH

Once the log-log graph has been completed, the estimator is in a position to perform countless calculations within the accuracy of the graph. The individual unit line is the most useful since it allows instant reading once the correct unit number has been selected. Unit numbers for individual lots are selected using the following rules:

1. If the lot selected includes the first unit, the unit number closest to the one-third point is used.
2. For all subsequent lots, the mid-point of the lot should be used.

The advantages of log-log plots compared with mathematical calculations are as follows:

1. Results are instant reading as noted above.
2. Allows quick response to requests for additional readings during discussion with management.

The disadvantages of log-log plots are as follows:

1. Coordinates are difficult to read, especially if values are large. Most results are merely approximations.
2. Inaccuracies resulting from the use of lot mid-points tend to underestimate costs.

As an example of the latter, consider the problem of determining the cost of units 51-750 in Figure One. The approximate mid-point is unit $400 and $95 is a close estimate for the lot average using the mid-point rule. The following calculations using time-sharing computer facilities indicates a more correct solution and points out the inaccuracy.

First Unit	Last Unit	Lot	Average Cost	Plot Point
51	750	700	$96.13	323.5
400	400	1	$93.09	400

Note that the true plot point for the block, calculated mathematically, is 323.5 rather than the lot mid-point of unit 400. If it were possible to accurately read the chart, use of the lot mid-point would have resulted in understating the cost by $3.13, an error of more than three per cent. This could be serious if the product being estimated were valued in thousands of dollars or more rather than hundreds. The lot mid-point is inaccurate only at the front of the curve when a large lot is considered. As you get farther out on the curve the lot mid-point becomes more accurate.

SELECTING THE PROPER SLOPE

Determination and plotting of the cost at the 250th unit or some other predetermined focal point may not be as difficult as the next step of determining the learning rate to be used in completing the projection. There are a number of considerations to be given serious thought before this step can be undertaken.

1. Has prior experience on similar components or processes been documented?

   As a cost accountant of more than thirty years' experience, I can readily indict the members of my profession because of the lack of cost experience data that is so prevalent throughout industry. Unfortunately, most cost accounting systems are tailored to the needs of the bookkeepers in determining the profit or loss and valuing the inventory for tax purposes and does not fit the needs of planners and cost estimators. Accurate cost data is a necessity. Too often the planner or estimator is forced to manufacture his own historical records. This can create problems in supporting Department of Defense proposals as the DCAA is looking for supporting detail which can be tied into the official company accounting system.

2. If accurate cost detail is available, do employees possess the knowledge to determine the slope resulting from use of the data?

   There are several good texts and training manuals which will train employees in the learning curve principles and show them how to measure results.

3. Will the production rate decline during the production phase reflected in the estimate?

   A serious drop in production rate will create an impediment to maintaining the learning rate determined from past experience. An increase in rate, on the other hand, should suggest consideration for increasing the learning rate, an important factor to be considered if competition for the order exists. There is nothing so depressing as an order lost through overcosting.

4. Did prior experience reflect technological improvements at a rate which will be difficult to maintain in the future?

   During the past ten years, many contractors have experienced fantastic learning rates because of the change from conventional machine tools to numerically controlled equipment. We do not attempt to reinvent the wheel. If the process improvements of the past have been included in the cost of the next order, it will be difficult to maintain the prior experience rate.

5. Is there an organized cost improvement program?

The machine operator contributes the least to the learning process. Many times he is limited by the feed and speed of the machine and improvements in his time are accomplished only by an increase in lot sizes to reduce setup costs or more attention to the amount of time he spends in the smoking area. Unless there is an organized program of cost reduction, involving value engineering, the likelihood of realizing a slope under 100% during the performance of the order does not exist.

6. How well were the prior jobs planned?

A poor job of planning will result in many production difficulties and high cost which, when measured for learning slope, will show excellent progress. A good job of prior planning will reduce initial starting costs and show poor learning but the final cost realized should be lower than that experienced with the poor planning performance.

7. Does prior year's experience reflect constant year's dollars?

In recent years inflation has become an important factor in cost estimating. The ideal situation is to estimate based upon current year's labor and expense rates and include a clause for inflation in the contract. This is not always possible, however. Cost experience should be analyzed to exclude the influence of inflation using factors included in the many indices used by the Bureau of Labor Statistics. If the contract is to include a clause covering inflation, these same indices can be used as a basis for future adjustments.

It is recommended that cost estimators prepare their estimates using learning rates which are free from the binds of inflation and leave the guessing of inflationary trends to the accountants or contract negotiators. There are several reasons for this. Basically, it is impossible to construct a constant slope curve with inflationary factors included which vary from year to year. Additionally, labor rates, with most companies, are negotiated with bargaining units at the corporate level and the estimator will not be in a position to predict how future year's negotiations will run. If it becomes necessary to reflect inflation in the estimate, it is recommended that the data be fed to the estimator from the proper corporate source and that these factors be added to the results derived from the use of the learning curve.

THE MULTIPLE PHASES OF PRODUCT LIFE

Cost estimators are faced with a decision when new products are to be manufactured in more than one phase; i.e., a development phase during which the product is manufactured using tools classified as temporary or general purpose, and a production phase in which the final configuration resulting from the development phase is to be manufactured using permanent special purpose tools.

There are four methods currently used in industry to reflect the change from the development phase to the production phase using the learning curve principles.

Figure Two shows the concept in which the production phase is considered to have no experience carried over from the development phase and the production curve starts at the top of the curve with the cost of the first production unit at unit number one. For this method to be realistic, the production phase must be considered completely divorced from the development phase from a tooling, facilities, and manpower basis. These conditions would be experienced only if the development phase was carried out in one plant or by a subcontractor and the production phase was carried out in another plant geographically distant from the first plant. It would also require that no development tooling be carried over into the production phase. The learning rates of the two curves would not, of necessity, have to agree. The development curve generally would not have as steep a slope as the production curve.

Figure Three illustrates the concept where the cost of the last development unit is considered the cost of the first production unit with the production phase starting over again at unit one. This concept is easily handled graphically by moving the cost of the last development unit laterally until it intersects the graph at unit one. From this point the production curve is drawn using the appropriate learning rate. This is a convenient method for estimating the production phase as it requires no production planning. There is nothing of an empirical nature, however, to substantiate the use of this method. The fact that it applied in one situation would not guarantee that it would withstand the test of time and apply in all situations. There is no substitute for production planning.

Figure Four illustrates the concept in which the cumulative development and production phases are continuous from a quantity-phased standpoint, but there is a step in going from the development to the production phase due to retooling. The magnitude of the step would be determined by the extent to which retooling takes place and facilities and manpower differ between the two phases. This is probably the most realistic of the various methods.

Figure Five shows a concept related to the previous one with the exception that it is considered that there will be starting costs on the first production units which will cause them to be more costly than the last development units.

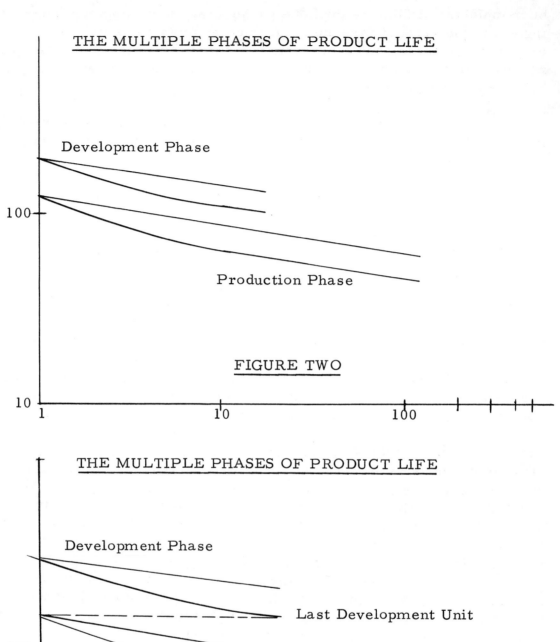

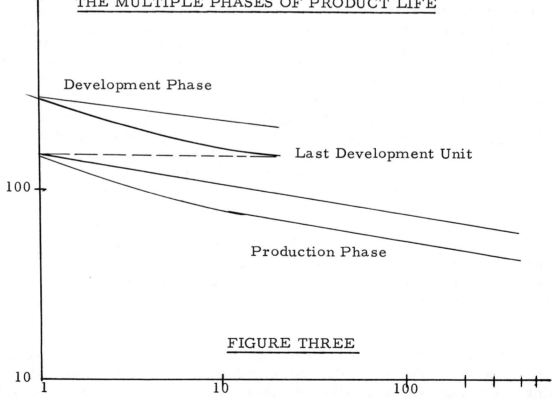

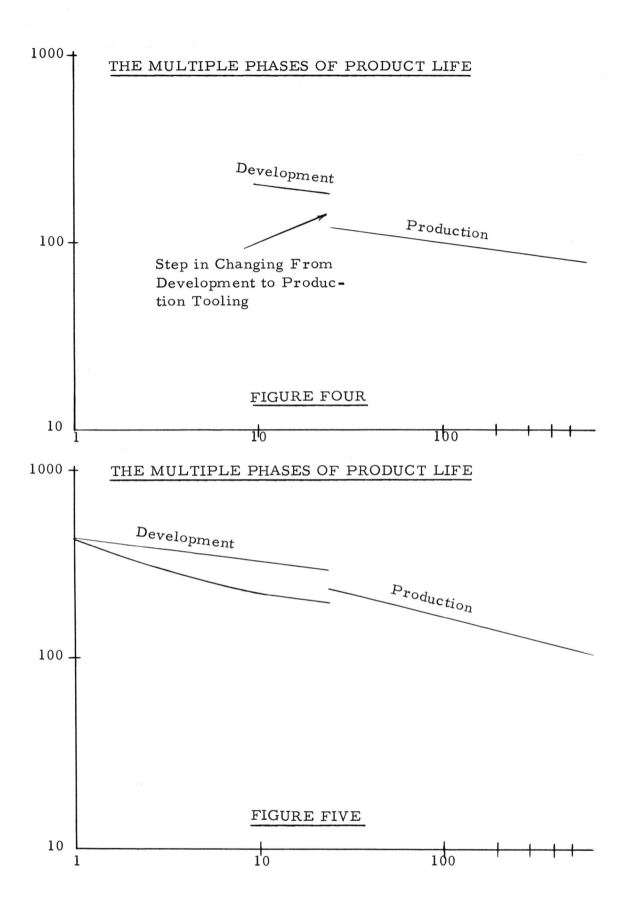

Which method is the correct one? Anyone of them can fit the situation under which you are operating and analysis is necessary in order to determine the characteristics of your situation.

THE CONSTANT SLOPE MYTH

A constant slope curve is theoretical only. Progress curves generally are "S" shaped over a long period of time. In the initial stages production and tooling problems, problems involved in meeting the performance guarantee, are a deterrent to learning and the slope is poor. As problems are corrected and production begins to flow on schedule, a cost improvement program can be introduced which will accelerate learning. The slope begins to steepen. As production moves a great distance out on the curve, a point of diminishing returns is reached. The slope begins to level off, become asymtotic, and in many cases can curve upwards. The customer many times is the greatest deterrent to cost progress but with reason. Many times cost reductions can be obtained only through major design changes which affect interchangeability with units in operation. Acceptance of the cost reduction would require provisioning of two different designs. Figure Six illustrates the general trend which follows the "S" shaped curve.

WHICH COST ELEMENTS SHOW LEARNING?

Although many users of the learning curve concept consider factory cost (the sum of material, labor, and burden) and sometimes selling price for learning curve applications, material, and labor alone are the only correct cost elements which should be applied. Material cost must be separated between raw materials and subcontracted items before applying theoretical or practical learning rates. In general, raw materials do not show very good learning. Subcontrac items, depending upon the extent to which the subcontractor carries on a cost reduction program, are a prime example for learning curve applications. The negotiating ability of both the customer and the supplier, however, will be a contributing factor. The use of learning curves by both sides will expedite the negotiations by providing common ground for discussion. Many companies have cost and price analysis units within the purchasing function which measure the performance of subcontractors in the performance of previous orders and prepare a calculation of the cost of the follow-on order based upon the perpetuation of the subcontractor's past performance. In such cases the contractor comes to the negotiation with a predetermined idea of what the subcontractor should charge.

If the price and cost analysis unit has only the past historical subcontractor selling prices as basic data, the projections lack the influence of future labor and overhead predictions which are influenced by bargaining unit negotiations and fluctuations in volume. Labor and overhead details will be available, however, under certain government contracts in which

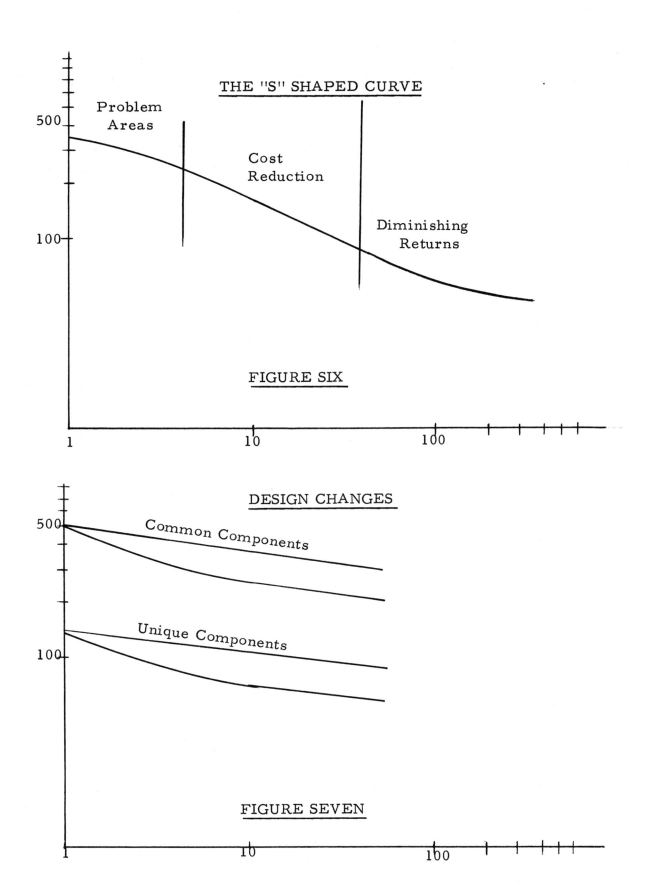

the supplier must provide a cost breakdown.

Direct labor projections should be on the basis of hours in order to allow flexibility in the application of labor rates. Overhead or burden is not a good application for the learning curve since this element varies in proportion to the overall business and is usually not related to a specific product. The application of the learning curve to material and labor separately will allow flexibility in applying varying rates of overhead in accordance with future forecasts.

HANDLING DESIGN CHANGES

One of the most frustrating experiences for a cost estimator takes place when an improved design is introduced in which many of the components are the same or similar to the original design. Treating the new design as a new curve starting over again at unit one is incorrect and dangerous since the components common to both designs will not learn as fast as the components unique to the new design.

If the design changes are minor, and there has not been a large quantity of the old design manufactured, it is the practice of some companies to start the curve over again, compensating for the slow learning of the common components by raising the slope; for example, from an 85% curve to an 87% curve.

Figure Seven illustrates an example in which the old and new design costs have been separated and individual curves have been drawn. This method is unwieldy, however, as it requires calculation of the position of the product on both curves and the addition of the results. If material and labor are handled as separate curves, the situation is compounded. One of the impediments to this method is that it cannot stand the test of time unless the cost system has been designed in such a way that the old and new design costs can be separated. Very few cost systems are so organized.

Figure Eight illustrates a more practical method. The cost of an individual unit or the average of the first lot of the new configuration is estimated. The cost is plotted on the experience curve of the previous design at a point equivalent to the total production of both designs. A line is then drawn laterally to the left until it intersects the old curve. This point represents the equivalent number of units of experience which are carried over from the old to the new curve. For example, if the 500th unit is the first unit manufactured to the new configuration, and projection of this cost laterally intersects the old curve at unit 350, there are 350 units of experience contained in the new curve. The first 100 units of the new design would be calculated as units 351-450 on the extrapolation of the old curve.

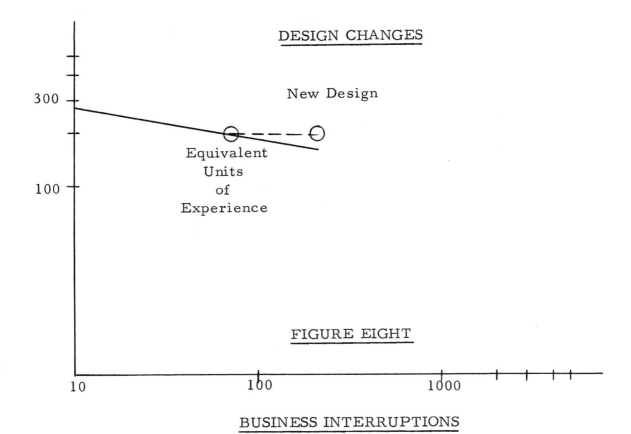

FIGURE EIGHT

BUSINESS INTERRUPTIONS

A gap in the production schedule exceeding six months can create a problem in the application of the learning curve to cost estimating. Before estimating the cost to be experienced after production resumes, the estimator must consider questions such as the following:

1. Has there been employee turnover so that experienced personnel are no longer available?
2. Are the tools still available or will replacement tools be necessary?
3. Are the facilities available or are they being used to capacity for other products thus requiring the use of less economical facilities and methods?
4. Has there been a change in the lot size compared with the previous production rate?
5. Are the previous vendors and subcontractors available? Is tooling available? Or will new vendors and new tooling be necessary?

If many of the answers to the above questions are negative, it will not be possible to start the curve at the point where the prior production ended. The method similar to the design change method illustrated in Figure Eight is useful. The objective is to determine the number of units of

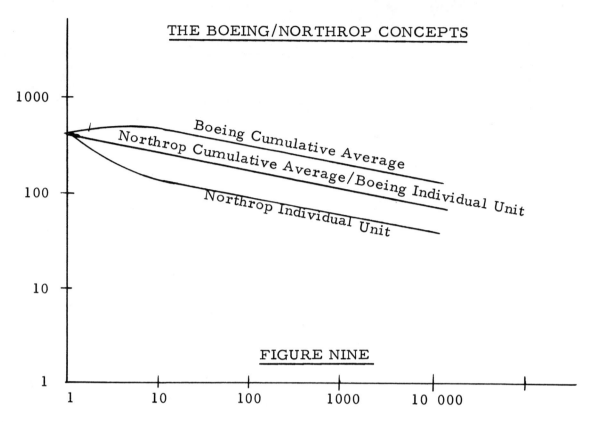

learning which have been lost by the interruption.

EXPLORING THE BOEING/NORTHROP PROBLEM

There has been much controversy over the interpretation of which curve is straight. T. P. Wright, in his generic article, interpreted the cumulative average line as straight and the individual unit line as curved. Boeing Aircraft has been given credit as the advocator of the reverse concept, while the Northrop Company has been named for the original concept. The government negotiators and the Defense Contract Audit Agency favor the Boeing concept. Figure Nine shows the two interpretations. Before the introduction of time-sharing computer programs, there were advantages to considering the individual average line straight since log-log graphs could be read instantly.

Different results are obtained with the two concepts if both curves start with the same value for unit one. The following shows a comparison using a value of 100 for unit one.

85% LEARNING CURVE

Unit Number	Northrop Concept Value	Unit Number	Boeing Concept Value
1	100.00	1	100.00
2	70.00	2	85.00
3	61.87	3	77.29
4	57.13	4	72.25
5	53.85	5	68.57
6	51.35	6	65.70
10	45.14	10	58.28
500	17.85	100	23.29
1000	15.17	1000	19.80

There are very few occasions, however, when the true value of unit number one is known. Most products are made in lots with the value of unit number one being a theoretical value calculated mathematically or by regressing on the log-log plot. Using time-sharing facilities which perform the calculation in seconds, the following comparison was made between the two systems using the value of the first lot of one hundred units as 100.

85% LEARNING

Northrop Concept		Boeing Concept		
Unit Number or Lot	Value	Unit Number or Lot	Value	Per Cent Difference
1	294.40	1	228.56	22.4%
2	206.08	2	194.27	5.7
3	182.16	3	176.65	5.4
4	168.18	4	165.13	1.8
5	158.49	5	156.71	1.1
6	151.18	6	150.15	0.7
10	132.96	10	133.20	0.2
500	52.50	500	53.23	1.4
1000	44.62	1000	45.25	1.4
1-50	117.65	1-50	116.62	0.9
51-100	82.35	51-100	83.38	1.3
51-150	77.57	51-150	78.57	1.3

Note that the Northrop cost for the first six units is higher than Boeing but the reverse is true when you reach unit ten and remains in that position. The difference between the systems appears to level off to roughly 1%. With this knowledge it should be possible for negotiations to be carried on under either system providing the focal point is other than the first unit.

CONCLUSIONS

Learning curves are an effective tool for use in estimating manufacturing costs. Their use, however, should not be undertaken without a thorough knowledge of the principles, problems, and pitfalls involved.

CHAPTER 5

TOOL ESTIMATING

American Machinist / Metalworking Manufacturing

Reprinted from American Machinist, September 4, 1961. © McGraw-Hill, Inc.

HOW TO ESTIMATE DIES, JIGS, and FIXTURES
FROM A PART PRINT

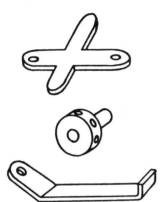

By Leonard Nelson, planning process engineer
Industrial Engineering Dept, Brown Instruments Division
Minneapolis-Honeywell Regulator Co

How long will it take—how much will it cost—to build a die, jig or fixture for a job?

An accurate estimate can be made in a matter of minutes while the job is still fresh off the drawing board, and before it has been released to production.

All that's needed is a part print, the tables in this article, and some knowledge of tool construction. Not only will this method determine the time and cost of tooling, it can determine which of several different methods of making tooling will be the fastest and most economical.

A small oversight or two won't impair the accuracy of the calculation, which is actually so simple that the burden of estimating can be turned over to an assistant.

Essentially, this method estimates tool cost in terms of hours, but it is easy to apply the shop rate to the time estimate to get probable tool cost for any given job. And, with a little more figuring (based on production rates, burden rate, and material costs), piece-part costs can be determined.

This shorthand guide to tooling estimation can be especially valuable to the shop man or engineer who must give on-the-spot estimates to part designers. In other words, this method is so fast and so accurate that the most economical and efficient part design can be determined while there is flexibility with respect as to how a part or product will be made—the modern way of attaining low overall costs.

How reliable *is* this estimating method? The data have not only been collected, checked, and rechecked over a period of years at Minneapolis-Honeywell, but have been used in other plants.

At first, actual tool building times may be on the high or low side compared to the estimate, but the deviations will be consistent. All that's necessary is to determine a suitable constant based upon *your* conditions.

COPYRIGHT 1961 BY McGRAW-HILL PUBLISHING CO, INC, 330 W 42nd ST, NY 36, NY

DIES, JIGS, and FIXTURES...

Section I — How to Estimate Dies

By looking at the part print and referring to the tabulated time data in this guide, you can estimate the hours required to build a "basic" die or a "combination" die. The word "basic" refers to a die that performs only one function. For example, a basic pierce hole die only pierces holes; a blanking die does only blanking. A combination die might pierce and blank or be a multiple-station progressive die.

Examples are provided for estimates of basic and combination dies, so that you can see how the tables are used.

Physical Limits of the Die
(Both Basic and Combination)

All time values given in the tables are based upon the construction of a die to these specifications:

1. The punch and die will be made of oil-hardening tool steel.
2. Punch and die blocks will be machined square, then ground before and after heat treatment.
3. These punch and die blocks will be mounted with screws and positioned with dowels.
4. The punch and die blocks will be assembled on the surface of the die shoe and punch-holder of the die set.
5. Time values are included in the estimate for die accessories such as gages, nests, strippers, punch pads, locating pins, spring pins, stops, knockouts, and pressure pads.
6. The minimum tolerance allowed for all dimensions will be ±0.002 in.
7. Material thickness will range from 0.010 to 0.250 in.
8. Physical limits of the die set will not exceed 1½ in. in thickness for die shoe and punch holder, and the assembling space will not exceed 6 in. in width by 8 in. in length.

Additional time is necessary under these circumstances:

1. Add 20% when high-carbon, high-chrome steel is used instead of oil-hardening tool steel. The extra time is required because of the difficulty encountered in the drilling, machining, and grinding.
2. Add 25% when the blank thickness is less than 0.010 in. More time is needed because the clearance between punch and die is extremely close, and extra care is required in the fitting, filing, and stoning of the punch and die.
3. Add 10% when the die set is milled to fit the die blocks. Usually, milling is employed when material thickness is greater than ¼ in., to insure stability of the die blocks.
4. Add 10% when the physical limits of the die set, as stated above, are exceeded. More time must be allowed to compensate for the extra handling and drilling of the die set.

Basic Hole Piercing Dies

Hole-piercing dies can be classified into three types, according to hole diameter.

Type	Size Range
1. Small	0.0156 - 0.1245
2. Medium	⅛ - ½
3. Bored	Over ½

The time values in Table 1 are

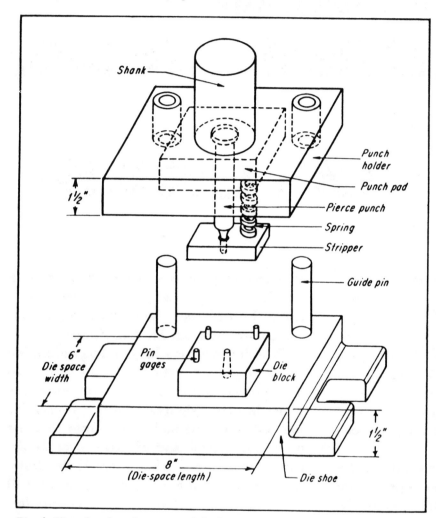

Fig. 1—Time data for a basic pierce hole die are given for a tool that does not exceed these physical limits. If the required die is larger, add 10% to compensate for extra handling that will always be involved

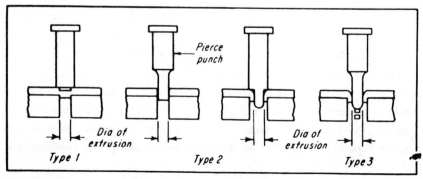

Fig. 2—Basic extrusion dies may push the metal part way out of the sheet, Type 1, or pierce a hole and extrude a collar or eyelet, Types 2 and 3

Calculations for Basic Piercing Dies (see Table 1 for basic data)

Example No. 1—To pierce five 0.031 in. dia holes in 0.009 in. stock. Oil-hardening tool steel (OHTS) die

```
Estimate:
  One basic die (small pierce hole) ...... 21
  Four additional holes (4 x 3.5) ........ 14
                                           35
  Material thickness allowance
    (25% of 35) .......................... 9
                                  total   44 hr
```
Example 1

Example No. 2—To pierce five 0.250 in. holes in 0.031 in. material. High-carbon high-chrome (HCHC) steel die

```
Estimate:
  One basic pierce-hole die .............. 16
  Four additional holes (4 x 2.5) ........ 10
                                           26
  Allowance for HCHC die
    (20% of 26) .......................... 5
                                  total   31 hr
```
Example 2

Example No. 3—To pierce five 0.745 in. holes in 0.188 in. stock. HCHC die, and extra-large die set.

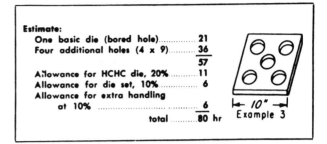
```
Estimate:
  One basic die (bored hole) ............. 21
  Four additional holes (4 x 9) .......... 36
                                           57
  Allowance for HCHC die, 20% ............ 11
  Allowance for die set, 10% ............. 6
  Allowance for extra handling
    at 10% ............................... 6
                                  total   80 hr
```
Example 3

Example No. 4—To pierce one 0.031 in. hole, one 0.250 in. hole and one 1.250 in. hole in 0.031 in. material. OHTS

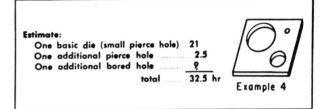
```
Estimate:
  One basic die (small pierce hole) ...... 21
  One additional pierce hole ............. 2.5
  One additional bored hole .............. 9
                                  total   32.5 hr
```
Example 4

based upon the diameter of the hole. Holes from a 1/64 to 1/8 in. are designated as "small pierce holes." Time values are greater than for medium pierce holes because of the extra care in drilling the die block, to avoid drill breakage, and the extra time for grinding and making the pierce punch. Holes with diameters of 1/64 - 3/64 in. require a special quill-type punch and holder.

Holes from 1/8 - 1/2 in. dia are designated as "pierce holes." Time values are based on the ease of drilling the die blocks, and the fact that standard drills are available.

Holes larger than 1/2 in. dia will be designated as "bored holes." Time values are based on the need to bore the hole in the die block, coupled with the extra time for grinding and machining the larger pierce punch.

Basic Blanking Dies

Blanking dies are estimated according to the periphery of the blank. The contour of the blank will be made up of one or a combination of the four different types of contours:

1. Straight lines
2. Angular lines
3. Curved lines that cannot be bored
4. Irregular lines

Time values for basic blanking dies are given in Table 2. Note that times

Table 1—Time Values for Basic Pierce Hole Dies
(Oil Hardening Tool Steel)

Hole Dia In.	Hr for Basic Die (one hole only)	Hr for Each Additional Hole
Small Holes		
0.0156 - 0.1245	21	3.5
Pierce Holes		
0.125 - 0.500	16	2.5
Bored Holes		
0.501 - 1.000	21	9.0
1.001 - 1.500	31	10.0
1.501 - 2.000	33	12.0
2.001 - 2.500	36	15.0
2.501 - 3.000	39	18.0
3.001 - 3.500	42	21.0
3.501 - 4.000	45	24.0
4.001 - 4.500	48	27.0
4.501 - 5.000	51	30.0
5.001 - 5.500	54	33.0
5.501 - 6.000	57	36.0

Extra Allowances:

If die is to be made of HCHC steel .... Add 20%

If material thickness is between 0.001 - 0.009 in .... Add 25%

If die set is milled to fit die blocks .... Add 10%

If die shoe and punch holder are over 1½ in. thick and die set greater than 6 x 8 in., for extra handling .... Add 10%

DIES, JIGS, and FIXTURES . . .

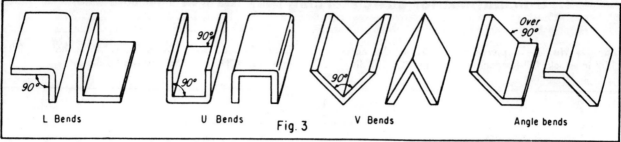

Fig. 3—Time values are given for basic form dies to make these four types of bends

Time Values for Three Types of Basic Dies

Table 2—Basic Blank Dies, Hr
(Oil-Hardening Tool Steel)

Periphery In.	Contour of Blank			
	Straight	Angular	Curved	Irregular
Up to 3	24	27	30	37.5
3 to 4	27	32	40	50.0
4 to 5	29	38	50	62.5
5 to 6*	32	45	60	75.0

*If periphery is greater than 6 in. use these values for every inch:

    Straight line 5.0
    Angular line 7.5
    Curved line 10.0
    Irregular line 12.5

Examples: Periphery 7 in., straight line: 7 × 5 = 35 hr
Periphery 7 in., curved line: 7 × 10 = 70 hr

When blank is a combination of contour lines:

I — Periphery is less than 6 in:
  Straight and angular lines —
  Sketch A, use "Angular" column
  Straight, angular and curved —
  Sketch B, use "Curved" column.

II — Periphery is greater than 6 in:
  Combination of contours — use combination-dies section
  Circular and straight lines —
  Sketch C, use "Straight" column.

Table 3—Basic Extruding Dies, Hr
(Oil-Hardening Tool Steel)

Dia of Extrusion	Type 1		Type 2		Type 3	
	Basic Die	Each Add'l	Basic Die	Each Add'l	Basic Die	Each Add'l
0.0156 to 0.1245	20	7.5	28	15.0	24	10.0
0.125 to 0.500	16	5.0	24	10.0	20	7.5
0.501 to 1.000	20	10.0	28	20.0	24	15.0

Table 4—Basic Form Dies, Single Bends Only, Hr
(Oil-Hardening Tool Steel)

Length of Bend	Width of Bend	90° Bends			Angle Bend
		L	U	V	
Up to 1	Up to 1	30	30	30	40
Over 1 to 2	Up to 1	32.5	32.5	32.5	43.5
Up to 1	Over 1 to 2	32.5	32.5	32.5	43.5
Over 1 to 2	Over 1 to 2	35	35	35	47
Over 2 to 3	Over 2 to 3	40	40	40	54

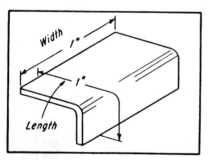

Fig. 4—Tabulated data are given in terms of variations from this "standard" size of 90° or L bend

are based on use of oil-hardening tool steel. Add 20% if high-carbon, high-chromium steel is selected.

Basic Extruding Dies

Basic extruding dies can be estimated according to the hole diameter of the extrusion and the type of extrusion. Consider three types of extrusions. (See Fig. 2):

1. An extrusion without a pierced hole.
2. An extrusion with a pierced hole requiring two stages for completion. First-stage piercing is followed by an extruding stage.
3. An extrusion with a pierced hole done in one stage.

Time values for basic extrusion dies are given in Table 3.

Basic Bending or Forming Dies

A basic forming die can be estimated according to the width, length and type of bend. Types of bends are classified in Fig. 3. Most forms are 90° bends. The L, U, and V forms are variations of the 90° bends. An angle bend is a bend other than 90°.

Time values are based on a standard form die for a part that contains a bend 1-in. wide by 1-in. long. See Fig. 4. All 90° bends that are less than 1-in. wide and less than 1-in. long will use the same value as the standard form die. For bends wider than 1-in. or longer than 1-in., an additional 2.5 hr per inch of width or length will be added to compensate for the additional machining, grinding, drilling, and assembling of larger form blocks. For angle bends, an additional 3.5 hr per inch of width or length will be added. Table 4 applies to one bend only of each type; if there are more than one, use Table 8.

How to calculate combination dies

For dies that perform more than one function, is necessary to estimate the time for each function and then add the figures obtained for the various functions. See the "Index for Time-Data Tables."

With these tables, it is possible to compare tool costs for different methods. See Example 1. Here two methods are compared.

- Method 1 employs a progressive die. The tool cost in terms of time is 133 hr.

- Method 2 involves use of three dies: pierce and blank, form two ears and the end, and final form. Here the tool-building time is 48.5 + 45 + 30 = 123.5 hr.

Thus, the progressive die requires only 9.5 hr additional to build and may well provide the lowest over-all piece-part cost.

Example 2 discusses a problem of piecing, blanking and forming that can be done by three methods. The second method has a distinct time advantage over the other two proposals.

Availability of time data therefore stimulates the estimater to consider several methods for making a part. And within a few minutes he is often able to show the cheapest method for making the part, so far as tooling cost is concerned, and to give a good guess of the actual piece-part cost. Such information is invaluable when the process engineer and the product designer are working together on manufacturing feasibility studies.

Time Values for Combination Dies (OHTS)
(Use these values for calculating the stations)

Table 5—Pierce Holes & Pilots in Combination Dies

Type of Hole	Hole Dia	Pierce Holes Estimated Times, Hr	Pilots
Small	0.0156-0.1245	7 hr/hole for the first five holes. For each additional hole thereafter add 3.5 hr/hole.	3.5
Pierce	0.125-0.500	5 hr/hole for the first five holes. For each additional hole thereafter add 2.5 hr/hole	2.5
Bored	0.500-1.000*	10 hr/hole for the first five holes. For each additional hole thereafter add 5.0 hr/hole.	5.0

*For holes greater than 1 in. dia, refer to Table 1, column marked, "Estimated Hr for Each Additional Hole".

Index to Time-Data Tables for Combination Dies

Type of Die	Use Tables			
	5	6	7	8
Pierce & Blank	X	X		
Pierce, Extrude & Blank	X	X	X	
Pierce, Extrude, Blank & Form	X	X	X	X
Pierce, Cutoff & Form	X	X		X
Pierce & Cutoff	X	X		
Form (more than one bend)				X
Pierce, Extrude & Cutoff	X	X	X	

Table 6—Blanking Stations in Combination Dies

Periphery	Contour of Blank			
	Straight	Angular	Curved	Irregular
Up to 3"	For odd-shaped blanks regardless of contour within blank, use a minimum of 15 hr/hole; if there are more than three similar holes, use 10 hr/hole thereafter.			
Over 3"—Add per inch of periphery	5.0	7.5	10.0	12.5

Table 7—Extrusions in Combination Dies

Extrusion Dia	Type 1	Type 2	Type 3
0.0156-0.1245	7.5	15.0	10.0
0.125 -0.500	5.0	10.0	7.5
0.501 -1.000	10.0	10.0	15.0

Table 8—Bends or Forms in Combination Dies

Length of Bend	Width	90° Bends	Angle Bends
Up to 1 in.	Up to 1 in.	15 hr/bend	20 hr/bend

Over 1 in. of length or width—Add for every 1 in., either length or width—2.5 hr for 90° bends and 3.5 hr for angle bends.

Additional Allowances	
HCHC	Add 20%
Thin Material	Add 25%
Mill Die Set	Add 10%
Extra Large Die Set	Add 10%

Calculations for Combination Dies

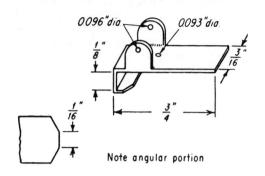

Note angular portion

Example 1—Made by Two Methods

Method 2—Three Dies

Pierce and Blank Die

Item	Reference Table	Calculation	Hr
3 small holes	Table 5	7 x 3	21
1 small pilot	Table 5	1 x 3.5	3.5
1½ in. of straight	Table 6	2 x 5	10
2 pierce holes (tabs)	Table 5	2 x 5	10
½ in. of angle	Table 6	½ x 7.5	4
			48.5

Form Die (2 ears and end)

3—90° bends	Table 8	3 x 15	45

Final Form Die

Basic form die	Table 4	1 x 30	30
		Total	123.5

Method 1—Progressive Die

Item	Reference Table	Calculation	Hr
3 small holes	Table 5	7 x 3	21
8 small pilots	Table 5	8 x 3.5	28
1½ straight outline	Table 6	2 x 5	10
2 pierce holes (tabs)	Table 5	2 x 5	10
½ in. of angle	Table 6	½ x 7.5	4
4—90° bends	Table 8	4 x 15	60
		Total	133

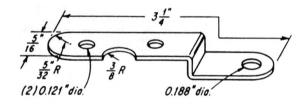

Example 2—Made by Three Methods

3 pierce holes	Table 5	3 x 5	15
½ in. of curve	Table 6	5	5
2—90° bends	Table 8	2 x 15	30
2 in. length	Table 8	2 x 2.5	5
			76

Method 3—Two Dies

Pierce and Blank Die

2 small holes	Table 5	2 x 7	14
2 small pilots	Table 5	2 x 3.5	7
3 pierce holes	Table 5	3 x 5	15
½ in. of curve	Table 6	5	5
6 in. of straight	Table 6	6 x 5	30
			71

Form Die

Basic die			
2—90° bends	Table 8	2 x 15	30
2 in. length	Table 8	2 x 2.5	5
			35

Total = 71 + 35 = 106

Example 2—Material 0.031 in. Brass

Item	Reference Table	Calculation	Hr
Method 1—Blank, Pierce and Form Die			
2 small holes	Table 5	2 x 7	14
4 small pilots	Table 5	4 x 3.5	14
3 pierce holes	Table 5	3 x 5	15
½ in. of curve	Table 6	½ x 10	5
6 in. of straight	Table 6	6 x 5	30
2—90° bends	Table 8	2 x 15	30
2 in. length	Table 8	2 x 2.5	5
			113
Method 2—Pierce, Cut-off, and Form			
2 small holes	Table 5	2 x 7	14
2 small pilots	Table 5	2 x 3.5	7

Section II — How to Estimate Jigs and Fixtures

This section is a "short hand" guide for determining the number of hours necessary for a toolmaker to build a jig or a fixture.

Jigs and fixtures are used to produce duplicate, interchangeable parts. This aim is accomplished by accurately locating and holding each piece in an identical position in the jig or fixture.

Time-determining factors in building a jig or a fixture are: size of the part to be held, its length, width and height. From the part print we determine the general appearance and shape of the work, the number of operations to be done, and the approximate dimensions of the jig or fixture.

Five basic components of a jig or a fixture are (See Fig. 5):

1. A baseplate or mounting plate— The baseplate is comparable to the die set in the die making. All fixture blocks are assembled on the baseplate by means of screws and dowels.

2. A locating means—The part is located accurately by pins, edge-locating blocks or nests, or V-blocks.

3. A clamping means — Several kinds of clamps are used: thumbscrews, knobs, quick-acting hand-operated clamps, and power-operated clamps.

4. A positioning means — Desired relationship between the cutting tool and a jig or fixture is established by a drill plate; in a fixture a set block would be the positioning means.

5. Tool guides or bushings— These are used in jigs for drilling and reaming holes accurately.

Classification of Jigs and Fixtures

As many as 40 or more types of fixtures will be used in a large plant. They are named according to the operation performed, for instance:

Assembly, staking, testing, locating, holding, positioning, alignment, drill, lathe, milling, reaming, tapping, grinding, turning, etc. For our purposes, let's put fixtures into four groups:

1. Drilling fixtures .... Drill press
2. Milling fixtures .... Milling machines
3. Bench fixtures .... Bending & burring
4. Turning fixtures .... Lathes, turret lathes

Time allowance to machine a purchased faceplate, to make a turning fixture, is 6 hr.

Physical Limits of Jigs and Fixtures

All tabulated time values are based upon construction of a jig or a fixture to the following specifications:

1. The steel employed for the baseplate is either hot or cold-rolled steel (HRS or CRS).

2. Locators, clamps, positioners, and tool guides (with the exception of the drill plate and supporting blocks) will be made of oil-hardening tool steel (OHTS).

3. All blocks will be machined square and ground.

4. All blocks will be positioned by dowels and fastened with screws.

5. Part-size limits will be 6 in. wide by 6 in. long, or 6 in. dia by 3 in. high, and these limits apply to parts that are flat, round or a combination of both and tubing.

6. Physical limits of the mounting plate will not exceed 1 in. thick by 12 in. wide by 12 in. long.

7. The minimum tolerance for jig boring will be ±0.0002 in.

Additional time allowances are provided when the jig or fixture exceeds the above physical limits. For instance:

1. An allowance of 10% is made when "standard" size limits of the baseplate are exceeded. The additional time compensates for extra handling and drilling of the plate.

2. An allowance of 25% is added when the standard physical limits of the part are exceeded.

3. An allowance of 10% is necessary when a trunnion jig or fixture is needed. More time must be allotted for extra handling and assembly of the intricate fixture.

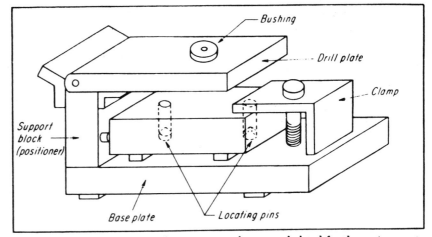

Fig. 5—Basic components of a fixture. Time values are tabulated for the various parts

Baseplates or Mounting Plates Fall in Three Categories

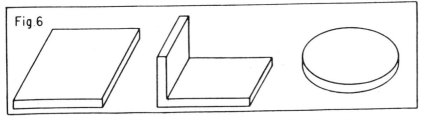

Fig. 6—Baseplates for drilling and milling fixtures are of various shapes

The selection of the proper mounting plate depends upon the "base" dimensions of the part, the general shape (rectangular, round, a combination of both, or tubing) and the type of machine to be used. The base dimensions of a part are the two linear bounds of the base of the part which locates itself on the mounting plate. Baseplates are customarily specified to be made of hot-rolled or cold-rolled steel plates of various shapes to suit the job.

Each baseplate of the four general types is different in construction. The size of the baseplates will be classified into small, medium and large:

A *small* baseplate will house a part with dimensions up to and including 1 x 1 in.

A *medium* baseplate will hold a part with dimensions greater than 1 x 1 in. and up to and including 3 x 3 in.

A *large* baseplate will encompass a part with dimensions greater than 3 x 3 in. and up to and including 6 x 6 in.

The baseplate will be larger than the part in base dimension; usually from 1 to 3 in. is allowed on each side. This space is needed to assemble the nest or the fixture blocks.

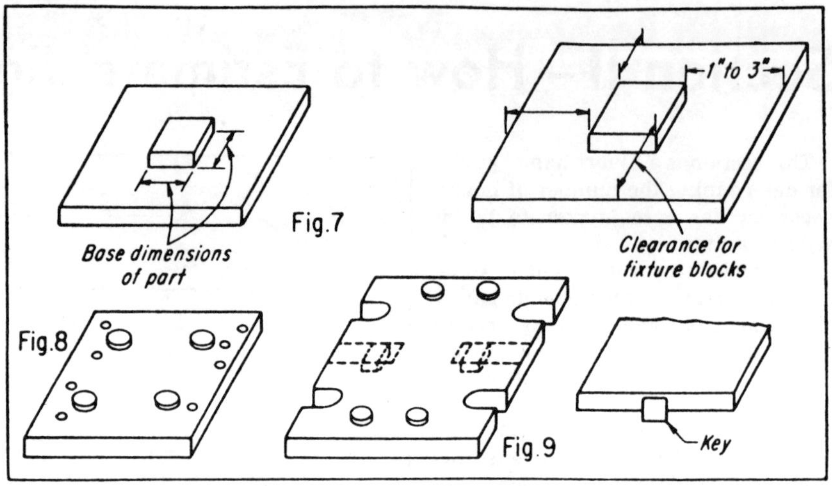

Fig. 7—In selecting a baseplate, two considerations are the linear bounds of the part (left) and clearance for fixture blocks (right). **Fig. 8—A baseplate** for a drilling fixture has four feet and rest buttons. **Fig. 9—Milling-fixture baseplates** require four rest buttons, two keyways and two keys

Table 9—Time Values for Flat Baseplates

Size	Thickness	Width	Length	Est. Hr.
		Drilling Fixtures		
Small	¾-1	3	3	6
Medium	¾-1	7	7	8
Large	¾-1	12	12	10
		Milling Fixtures		
Small	⅝-1⅛	3	6	10
Medium	⅝-1⅛	6	9	12
Large	⅝-1⅛	10	14	14
		Bench Fixtures		
Small	⅜-⅝	4	6	5
Medium		6	8	7
Large		9	12	9
		Faceplates		
Large	Machining purchased plate			6

Notes:

Time Values Include:

1. For drilling, milling and bench fixtures—cutoff, squaring, grinding, layout, plus drilling, reaming and tapping for locators, clamps form blocks or vise grip as required, and assembly.

Additionally:

2. For drilling-fixture baseplates—making four feet for the baseplate and four rest buttons for the work.
3. For milling fixture baseplates—milling four bolt slots and two keyways, making and fitting two keys, and making four rest buttons.
4. For bench-fixture baseplates—making the vise grip.

Turning fixture—

Time values include (a) drilling, reaming and tapping the *purchased* faceplate for installation of locators, clamps, set blocks, (b) providing bolt holes to fasten faceplate to chuck, (c) milling slots for guiding clamps, and (d) assembly of faceplate to lathe chuck.

Baseplates for Bench Fixtures (Tube Bending)

Many bench fixtures are used for bending tubing. In order to determine the baseplate size of a bending fixture, the outline of the completed part is the key. The width and length of the bend are comparable to the base dimensions, as shown in Fig. 10.

A small baseplate will house a tubing part up to and including 1 in. wide by 4 in. long. A medium baseplate will house a tubing part up to and including 3 in. wide by 6 in. long. A large baseplate will house a tubing part up to and including 6 in. wide by 10 in. long.

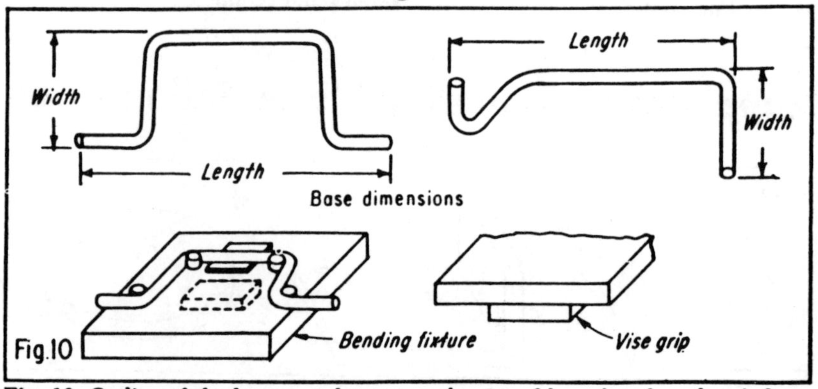

Fig. 10—Outline of the bent part determines the size of baseplate for a bench fixture

Angular Baseplates for Circumferential Operations

Angular baseplates are used in conjunction with a regular baseplate when the part is round and requires work done on its circumference. The shape of the part determines use of an angular baseplate in drilling, milling and turning fixtures.

Time values (Table 10) are based upon the use of the angle plate as a base or in combination with another baseplate. The C-shaped angle plate is also used in the making of a trunnion fixture. Time values include the complete machining of the angle plate along with the drilling, reaming and tapping for locators, clamps and set blocks.

Fig. 11—Angular baseplates are mounted on flat baseplates to hold round parts requiring work on their circumference. Fig. 12—Plain and C-shaped types, are given time values for three sizes, Table 10

Table 10—Time Values for Angular Baseplates

	Size	Angle as Base	In Combination	C Angle
		Hr	Hr	Hr
Small	3 x 3 x 3	10	5	7
Medium	6 x 6 x 6	15	9	11
Large	6 x 9 x 9	20	13	15

Time values include complete machining of the baseplate, plus drilling, reaming, and tapping for locators, clamps and set blocks.
"In combination" refers to use of an angle plate with a flat baseplate.

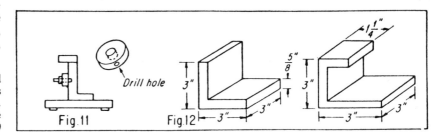

Special Baseplates for Two or More Operations

Baseplates discussed so far are used for one operation—in one plane. When two or more operations are required on a part, they are ordinarily done by two separate fixtures. With a special fixture, however, the two operations can be performed in the same fixture with minimum of movement on the part of the operator. Examples of such fixtures are:

1. **A box or tumble jig** is used for drilling opposed holes. The support legs are of such length that all holes of the same diameter can be drilled with the same spindle stop setting. Both holes can be drilled with only

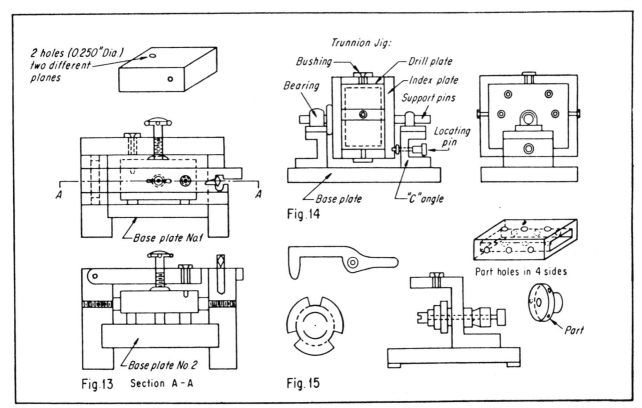

Fig. 13—Box or tumble jigs incorporate two baseplates

Fig. 14—Trunnion jigs are usually employed only for large parts

Fig. 15—Index drill jigs require 25 to 40 hr to make

DIES, JIGS, and FIXTURES . . .

one positioning for depth, which is made during setup. In a box jig, two baseplates are necessary. See Table 11 for time values.

2. **A trunnion-mounted jig or fixture** is used to drill opposed holes in two or more planes. The device is mounted between two pins and rotated by means of an index plate. The time to make the baseplate (12 in. wide x 18 in. long) is approximately 14 hr.

Trunnion jigs are usually associated with large parts. It doesn't pay to build this type of jig for small parts.

Table 11—Time Values for 2 Baseplates for Box Jigs

Size of Part	Base Thickness	Base Dimensions Width Length	Est. Hr.
Small	¼-1	3 x 3	11
Medium	¼-1	7 x 7	14
Large	¼-1	12 x 12	17

3. **An index drill jig.** When the part to be drilled is round and more than one hole is to be drilled on the circumference, an index drill jig is used. The device is mounted on an angle plate and uses an index assembly to rotate the part. The index assembly consists of a lever, index hub, clamp and locating bar, clamp knob and washer.

The time value for the making of the index assembly will vary between 25 and 40 hr, depending upon the complexity of the assembly.

Five Locators for Baseplate Use

Every part has to be located on the baseplate to insure identical parts. A minimum of two-point location is necessary. There are many types of locators:

1. Pin
2. Block
3. Edge block
4. Nest
5. V-block

PIN LOCATORS — Parts with machined holes can use pin locators effectively—either straight pin or diamond pin. Small pin locators are up to ½ in. dia by 2 in. long; large pin locators, 1½ in. dia up to 2 in. long. Time values are given in Table 12.

BLOCK LOCATORS FOR TUBING— Block locators are used in bench fixtures, especially for bending. Time values include the cutoff, squaring, grinding, drilling and assembly of these blocks on the fixture.

	Est. Hr.
Small —	2.5
Medium —	3.5
Large —	4.5

EDGE BLOCKS AND NESTS—Edge and nest blocks are an effective means for locating a flat sheet metal part. Time values include the cutoff, squaring, grinding and assembling the blocks. The estimated time for one nest or edge is 2.5 hr.

V-BLOCK LOCATORS—These are an effective means of locating round parts. Time values include the cutoff, squaring, grinding and assembling the blocks.

	Est. Hr.
Small —	3.5
Medium —	4.5
Large —	5.5

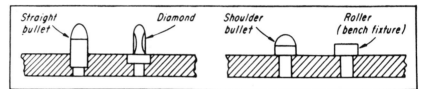

Fig. 16—Pin locators are used to position parts having machined holes

Table 12—Time Values for Pin Locators, Hr

Type	Machining	Jig boring	Assembly	Total
Small—up to ½ x 2 in.				
Straight	0.25	1.00	0.50	1.75
Shoulder or roller	0.50	1.00	0.50	2.00
Diamond	0.50	1.00	0.50	2.00
Large—½ to 1½ in. dia, to 2 in. long				
Straight	0.50	1.00	0.50	2.00
Shoulder or roller	0.75	1.00	0.50	2.25
Diamond	1.00	1.00	0.50	2.50

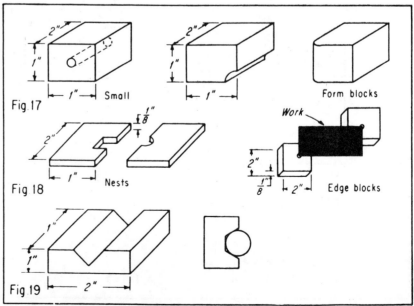

Fig. 17—Edge blocks and form blocks for bench fixtures require from 2.5 to 4.5 hr to make, depending on size. Fig. 18—Nests and edge blocks locate flat parts. Fig. 19—Round parts are customarily located in V-blocks

Positioners Include Support and Set Blocks

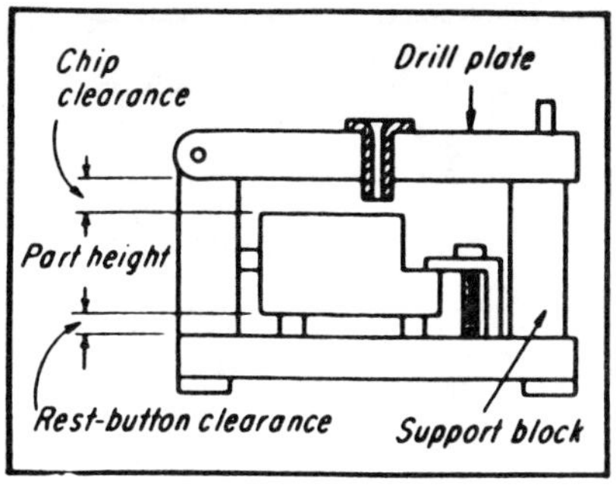

Fig. 20—Support blocks position workpieces in fixtures, and drill plates position and guide the tools

Table 13—Time Values for Drill Plates and Support Blocks

Part		Est. Hr.	2 Drill Plates for Box Jig	Index Plate Trunnion Jig
Small	⅝ x 3 x 3	3	5	
Medium	⅝ x 7 x 7	5	9	
Large	⅝ x 12 x 12	7	12	8

Support Blocks

Height of part, In.			Est. Hr.
1	Small	⅝ x 1½ x 3	2
3	Medium	⅝ x 3½ x 7	3
5	Large	⅝ x 5½ x 12	4

Time values for drill plates and support blocks include cutoff, squaring, drilling and assembly on the fixture.

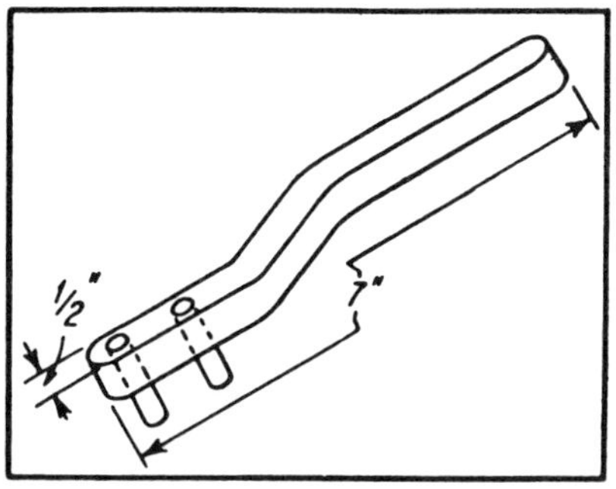

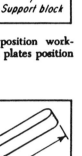

Fig. 21—Bending levers require 3 hr to make and assemble to the fixture

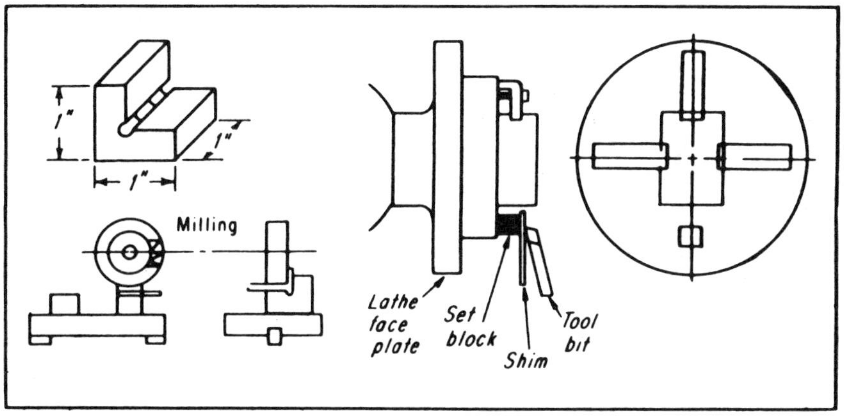

Fig. 22—Set blocks are used for positioning a milling cutter or a toolbit

Drill Plates—The size of the drill plate is roughly the size of the baseplate. The support block will be the width of the baseplate and the height will be approximately ½ in. more than the height of the part. The ½ in. space is necessary for chip clearance and room for the work to be supported on rest buttons. Therefore, the support-block dimension will depend upon the height of the part.

Index plates are similar to drill plates. They are used on trunnion jigs for large parts only.

Set Blocks—In the making of a milling or turning fixture, a set block is employed for positioning the milling cutter or the toolbit. The time value for a set block is 2.5 hr.

Bending Levers—In the making of a bending fixture, the most common type of bending lever (Fig. 21) requires 3 hr for cutoff, squaring, grinding, drilling, locating, bending and assembly. This device is shown as being 7-in. long.

Calculation for a Typical Fixture

Example—To estimate a milling fixture for a part with base dimensions of 6 x 6 in. This is a *large* part.

Estimate—Time values are listed for the various fixture elements that the estimator deems essential to hold the part shown.

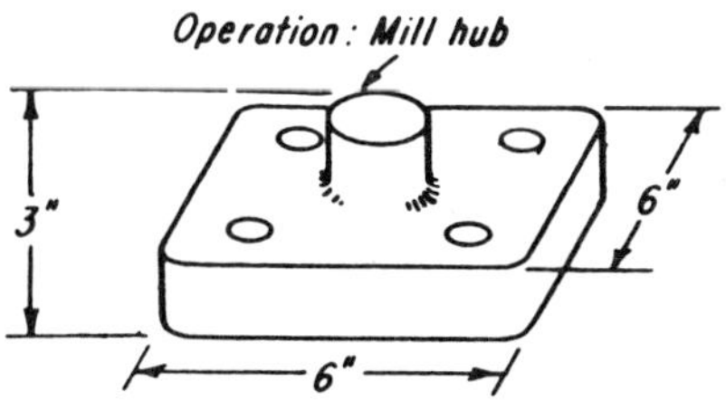

Item	Reference	Est. Hr.
Baseplate	Table 9	14
Locators		
one shoulder	Table 12	2
one diamond	Table 12	2
Clamps	page 122	16
2 guided		
Positioners		
Set block	page 121	2.5
Jig bore & inspection	Table 16	5
		Total 41.5

DIES, JIGS, and FIXTURES...

Clamps Require Time Allowances

There are many types of clamps used in the making of jigs and fixtures. Time data are given for the four most common clamps.
1. Hand knob clamps
2. Strap clamps with heel
3. Strap clamps with special heel and pin for guiding on turning fixtures
4. Quick-acting milling clamps

Knobs and thumbscrews are stock items. See Table 14 for time allowances for drilling and assembly of fixture part.

For strap clamps with heel, time values include cutoff, squaring, slotting, grinding and assembly of the strap clamp and heel block:

    Small — 3 hr.
    Medium — 7 hr.
    Large — 5 hr.

Guided strap clamps have a heel and pin for easy movement in the milled slots of milling or turning fixtures. Time values are:

    Small — 6 hr.
    Medium — 7 hr.
    Large — 8 hr.

Quick-acting milling clamps are purchased. Allow 1.0 hr for assembly.

Table 14—Time Values for Hand Knobs

Size of Part	Plain	With Block	Thumb Screw
Small	1.5	2.5	1.0
Medium	1.5	2.5	1.0
Large	1.5	2.5	1.0

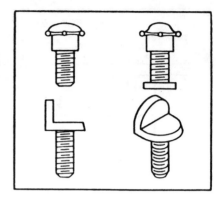

Fig. 23—Time for drilling and assembly of fixture parts must be considered, when knobs and thumbscrews are used

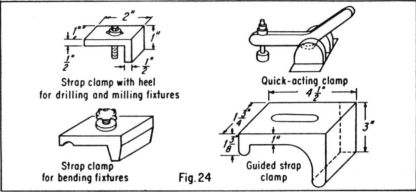

Fig. 24—These four types of clamps are commonly used in making jigs and fixtures

Guide Bushings

Table 15—Time Values for Guide Bushings (OHTS)

Bushing Dia.	No. of holes	Shoulder type	Headless type	Jig boring & assembly
up to ⅜ in.	1	2	1.5	0.5 hr/hole
	2 to 4	1.5/bushing	1.0/bushing	
	5 or more	0.75/bushing	0.5/bushing	

Time values are assigned for jig boring the hole for each drill bushing. Separate time values are necessary for setup and layout on the jig boring machine. This value is influenced by the size of the part and the type of jig.

A drill plate is jig bored for press-fit shoulder bushings; an index plate is jig bored for headless bushings.

Reprints of this and other AM Special Reports are available from Reader Service Dept., American Machinist, 1221 Ave. of the Americas, New York, N.Y. 10020. Single copies of this and most other reports are $1 (prices are reduced for quantities over 24).

Jig Boring—Setup and Inspection

Additional time must be allowed for the setting up the jig or fixture on the jig-boring machine for boring the drill plate, index plate and the baseplate (for the accurate position of the locators). At the same time the jig borer inspects the relative parts of the fixture. The time allowance is greater for special jigs, box and trunnion, because of the complexity.

Table 16—Time Values for Jig-Borer Setup and Inspection

Part Size	Est. Hr.
Regular Fixtures	
Small	3
Medium	4
Large	5
Special Fixtures	
Box or tumble	6
Trunnion	8

Full-cost estimating practices for manufactured products require recovery of the costs for direct labor, material, overhead, general and administrative, and other nonrecurring initial costs. Permanent tooling is a front-end cost that impacts the estimated cost and price.

PHILLIP F. OSTWALD and PATRICK J. TOOLE
University of Colorado, Boulder, Colorado

Figure 1. This fixture for a boring mill is an example of permanent tooling.

How do you handle tooling costs?

Permanent tooling consists of special equipment required in adapting machinery to a specific production process. It can take the form of jigs, fixtures, molds, dies, etc. A sample of a fixture is shown in Figure 1. Permanent tooling should not be confused with perishable tooling, such as drill bits and milling cutters, for which usage incurs a direct operational expense and is therefore included in overhead cost. Permanent tooling is a capital asset; its depreciation is subject to IRS regulations.

The problem of cost estimating permanent tooling as one element of the full cost of a product is often overlooked or over-simplified. Firms fail to examine their tooling write-off policies and the procedures that their estimating departments use. Tool write-off is inherent in pricing. As a concealed policy, its impact is important and needs to be questioned.

A survey of a diversified list of 51 manufacturing companies revealed that 17.6 percent do not estimate tooling cost at all. Reasons given for not estimating were "lack of time," "no information," or "don't know how." Of those companies in which tooling cost is estimated, 56 percent use a simple overhead rate method. These statistics indicate the lack of understanding that many companies have in dealing with amortization and estimating of tooling costs.

There are good reasons why tooling cost should be treated differently from other costs, such as plant and equipment, royalties, R & D, and setup costs. Most companies maintain an integral tooling department that can be discontinued under adverse conditions;

197

COST ESTIMATE AND PRICING SUMMARY

Estimated Cost of __Arthritis-Hand Exercise Unit__
Part No. __X 60371__   Customer's Drawing No. __492-7582__
Quantity __15,000 per year, 1,200/month__   Customer __A.B.C. Company__

Direct Material	$	1.2482
Freight In 2.51%		0.0313

Pack __15 per 18" x 24" x 12" box__   Total Material   1.2795
Weight __3.41 pounds__   Scrap   3.4%   0.0435
                                      Total $ 1.323

Department	Direct Labor	Variable Burden %	Variable Burden Amount	
4	$ 0.1135	168.8	$ 0.1917	0.3052
3	0.0726	132.2	0.0960	0.1686
7	0.0216	146.1	0.0316	0.0532
2	0.0731	107.9	0.0789	0.1521
11	0.7440	177.9	1.3235	2.0675

Cost to Manufacture:	$	4.0696
Contribution:		27%
General & Administrative:		7.2%
Commissions:		1.0%
Freight Out:		--
Selling Price:	$	6.25

Tooling	$ 46,200
Outside	6,000
Total	52,200
10% Profit	5,220
Total	$ 57,420

Figure 2. Example of a cost and pricing summary. Note, the tooling is listed as a separate charge paid directly by the customer.

therefore, it qualifies partly as a variable operational expense. Tooling is crucial to one product or product line while plant equipment is usually tied to multiproducts. Recovery of tooling cost is a discretionary policy of the firm. Finally, setup charges occur with each production run, but tooling cost results only from the initial decision to manufacture the product.

Magnitude of the problem

The problem and its magnitude depend on the kind of product and type of customer. Over-the-counter consumer-type products, OEM (original equipment manufacturers), one-client or sole-source, large distributors, and the government and its many departments and agencies call for various approaches to estimating and cost amortization.

Tooling cost is not a problem for the 6,600 independent tool and die shops that design, construct and sell only tooling. The estimating of tooling costs by tool and die shops does not involve the problem of capital recovery since ownership of the tool transfers to the buying company.

Yet, there are some 300,000 firms that manufacture products for sale, and a substantial majority of these firms use permanent tooling. Many companies have a tool and die department and are able to produce their own tooling either partially or fully, but this does not simplify the cost estimation or amortization problem.

By 1980, tool and die industry sales should reach $4.5 billion nationally, an average growth rate of 7.3 percent. Captive shops should produce ten times that figure. It is surprising that little attention has been given to the estimation of costs of this collective magnitude.

Estimating tooling

The method of estimating tooling cost depends on the available information. If a tool drawing is unavailable, an experienced tool estimator uses the component design and the operation process sheet to provide a 'guesstimate' of time or dollars. When a tooling design is available, a bill of materials and standard time data are used. The time is estimated in "units," which in the tool estimating parlance means hours. The hours are extended by the cost for labor and other indirect charges for the tool and die department. Developing a full cost allows for a make-or-buy comparison with an outside tool shop.

Some companies follow the practice of using an outside quotation to estimate their internal costs. This practice is used whenever the product company does not have the information or resources to do its own estimating. One company has used regression equations to estimate a class of permanent tooling, (see Klein, R. S., Tair, H. J., "Fatter, better tooling estimates," *Industrial Engineering*, December 1971).

Direct sale of tooling

The case in which the firm sells the tooling with the product deserves special attention. This method requires that the firm quote its tool cost separately from the unit or total product cost. Forty-eight percent of the 47 companies responding to the survey separated tooling charges from the primary product estimate. The usual practice is illustrated by the example in Figure 2. The total tooling related cost of $57,420 is quoted separately from the unit selling price of $6.25.

After the sale is made, the customer holds title to the tool, but the tool may be stored with the original supplier and not with the owner. A storage cost is sometimes charged the customer-owner. If the original tool builder desires to use the tool on other parts, special permission from the new owner is ordinarily required. If the "borrowed" tooling is damaged cost for refurbishment is paid by the user.

Those firms that buy the tooling view it as a means to maintain control, and they are free to move the tool to another supplier. Transportation costs are borne by the owner. Companies that practice "sourcing" move tools shop-to-shop as competition dictates. Transfer charges, which are not always nominal, are paid by the owner.

Amortizing tooling costs

The direct costs involved in the design and manufacture of tooling are the most significant capital outlays. In general, longer production runs provide for a greater "spread" of tooling costs, but require more expensive tooling. Tool repair may become a significant item if the production period is extended. After a decision has been reached on the type of tooling to be used, and after a cost estimate has been derived for the tooling, a number of techniques can be used to amortize the tool cost for the product-cost estimate.

Technique number one: Tooling cost amortized over a fixed product quantity. In this method a decision is made concerning the quantity (Q) for the cost estimates, to "recover" the tooling cost. Q must be the realized product quantity and not the component quantity. Significant error will result if tooling capitalization per item is underestimated by overestimating Q.

Q can be determined by an operational rule such as the quantity generated by a fixed number of production runs, by a time constraint such as one or more years of production, or by a more complicated mechanism such as the quantity expected to yield maximum profit.

A conservative policy is to include the cost of tooling over the first lot, year, or model. Fear of obsolescence of tooling due to design changes, competition reaction, etc., is the prevailing attitude. Out of a survey of 40 companies covering a spectrum of size and revenue, 50 percent amortized over a lot quantity less than a model or calendar year, 12 percent used the first year's quantity, and 38 percent used more than one year's quantity.

The conservative policy has been abused under certain circumstances. In auditing estimates, we have found that if the practice was to amortize tooling over say, the first lot, subsequent lots also carried the charge. The cost was overlooked in the follow-up estimate and later costs were overstated.

The tooling cost estimate is amortized by the simple formula:

$$\frac{\text{Total tooling cost}}{Q}$$

Figure 3 summarizes this technique. The original amortization number was set at 15,000 with a selling price of $7.50/item. This produced a profit loss of $0.40 per unit. Actual production figures are varied from 15,000 to 60,000 to show the effect of realized production and amortization on net profit per item. It is obvious from this example that the quantity Q has a dramatic effect on profit.

Technique number two: Tooling cost amortized over time and quantity. Including the time-value-of-money in the amortization is not a standard practice in cost estimating. However, if a long period of time is involved in production and marketing, cost estimates, profits, and production decisions can be affected. The selling price of the item and true variable expenditures such as labor costs can be assumed to increase proportionally to the rate of inflation. Using fixed time values for these estimates is an acceptable practice. In contrast, tooling costs are incurred at the beginning of the production period and may be time amortized to give a more balanced perspective to the cost estimate. For the example in Figure 3, assume a continuous discount rate of $\alpha = 0.06$. This rate reflects the time value of money and does not include capital recovery concepts. The tooling cost/unit is calculated by the formula:

$$\left(\frac{TC}{Q_n}\right) e^{-n\alpha}$$

where:
TC = $57,420
Q_n = 15,000
n = Number of years (1, 2, 3, and 4)

COST ESTIMATE AND PRICING SUMMARY

Product Name _____

Basis for Prices

Distribution [X]   OEM [ ]   Other _____

Quantity	15,000	30,000	45,000	60,000
Selling Price	7.50	7.50	7.50	7.50
Estimated Variable Costs (1)	4.026	3.8734	3.7708	3.7123
Standard Variable Profit	3.474	3.6266	3.7292	3.7877
less scrap and rework	0.0435	0.0435	0.0435	0.0435
Manufacturing Variable Profit	3.4305	3.5831	3.6857	3.7442
less out of pocket costs (2)	3.828	1.914	1.276	0.957
Adjusted Mfg. Variable Profit	− 0.3975	1.6691	2.4097	2.7872
less other variable costs (3)	--	--	--	--
Net Variable Profit	− 0.3975	1.6691	2.4097	2.7872

Estimated Variable Costs (1)

Direct Material	1.2795	1.3030	1.2967	1.2967
Direct Labor	1.0248	0.9736	0.9443	0.9255
Variable Burden	1.7217	1.5968	1.5298	1.4901
Total	4.026	3.8734	3.7708	3.7123

Out-of-Pockets Costs	Estimate	Product Labor Costs	Hours	Rate	Estimate
Tools and Fixtures	$ 57,420	Design	-	-	-
Equipment	--	Drafting	-	-	-
Outside Design	--	Development	-	-	-
Extraordinary Marketing Costs	--	Specification	-	-	-
Outside Sample Costs	--	Manuf. Engineering	-	-	-
		Quality Control	-	-	-
Total	$ 57,420	Technicians	-	-	-
Less Advance by Customer	--	Model Shop	-	-	-
Adjusted Total	$ 57,420	Total	-	-	-

$\frac{\text{Out-of-Pocket Costs}}{\text{Quantity}}$ = $ 57,420 ÷ 15,000 = $ 3.828 Per Unit (2)

Other Variable Costs (3) - None

Figure 3. In this cost estimate and pricing summary, tooling cost is amortized over four quantities with a fixed price of $7.50. A $.39 loss per item is incurred with a run of only 15,000. This summary shows the dramatic effect the quantity produced has on profit.

Amortized Net Profit

Net Profit/Item

n	Q_n	Quantity Only	Time and Quantity
1	15,000	− 0.3975	− 0.1746
2	30,000	1.6691	1.8855
3	45,000	2.4097	2.6199
4	60,000	2.7872	2.9914

Figure 4. The effect of quantity and time on amortized net profits.

Figure 4 compares the effect of quantity and time on amortized net profit per item. For example, if $n = 3$, then the time amortized tooling cost is:

$$\left(\frac{57,420}{45,000}\right) e^{-.18} = 1.0658$$

and the net profit (from Figure 3) changes to $2.4097 - (1.0658 - 1.2760) = 2.6199$ (or $3.6857 - 1.0658 = 2.6199$).

Technique number three: Tooling cost amortized to match tax write-offs. A complicated scheme, that few companies employ, matches the write-off of tooling to expenses as reported for taxes. There are contradictory notions and mistakes that attend this method. It is erroneously believed that accelerated methods of tax depreciation provide the limiting amount that a company can recover in sales through full-cost estimating. There are variations to this policy naturally, but the key is the matching of expected capital-recovery tooling income (as seen through sales revenue) versus acceptable tooling-tax expenses.

Technique number four: Avoiding amortization by use of a fixed overhead rate. In the survey, 56 percent of the companies treated tooling as overhead. This approach assigns tooling cost to the product estimate by the use of a full absorption overhead rate. Tool estimating and amortization are not required. The policy is easy to use, minimizes time required for estimates and is probably acceptable when tooling costs account for only a small portion of total costs. However, indiscriminant use of this method may distort the product estimate by under- or overstating the tooling cost for the product.

Other practices which are improperly labeled amortization policies deal with the discussion to: construct the tool internally or buy it from a tool and die shop or a combination of the alternatives: and use formulas to determine the amount of money or quantity before tooling is authorized. Neither of these two points bear on the problem of the amortization, although they are important operational questions.

IRS-requirements

The depreciation regulations of the Internal Revenue Service govern the amount of permanent tooling cost which may be deducted in a given tax year. The investment in permanent tooling qualifies for additional first-year bonus depreciation of 20 percent of cost if the useful life is at least 6 years. Alternately, the straight-line method may be used for depreciating the remaining cost, so an accelerated method may be used if the useful life is three years or more.

The company may determine appropriate and reasonable useful lives for income tax depreciation, or may utilize the Class Life Asset Depreciation Range System (ADR) of the IRS. ADR is based on broad industry classes of assets, and it offers a range of years for each class from which a depreciation period is selected. For many industry classes, a specific range of years is stated for permanent tooling. For example, the range is three to five years for special tools used in making fabricated metal products. For some industries, only one asset guideline class covers all equipment including tooling. An example is the manufacture of leather, which is provided a range of 9 to 13 years. The important observation is that the allowable ranges are significantly shorter for permanent tooling than other equipment for which the IRS has provided special tooling classes.

If the ADR system is used, the firm does not have to justify its retirement and replacement policies. Once a depreciation period is selected for an asset under ADR it cannot be changed during the remaining period of use of the asset. The ADR election is annual, and it applies to assets first placed in service after 1970.

Bibliography

Ostwald, Phillip F., *Cost Estimating for Engineering and Management*, Prentice-Hall, Inc., Englewood Cliffs, New Jersey, 1974, 493 pp.

Nelson, Leonard, "How to Estimate Dies, Jigs, and Fixtures from a Part Print," *American Machinist/Metalworking Manufacture*, September 4, 1961, Special Report No. 510, McGraw-Hill, New York, New York. IE

Phillip F. Ostwald is a Professor in the Department of Engineering Design and Economic Evaluation at the University of Colorado where he has been active in teaching and research, primarily in cost estimating. Also, he regularly teaches short courses in manufacturing cost estimating for the American Management Association. Dr. Ostwald holds a BSME from the University of Nebraska, an MSIE from Ohio State University, and a PhD from Oklahoma State University. He is a Senior Member of AIIE.

Patrick J. Toole is an Assistant Professor in the Department of Engineering Design and Economic Evaluation at the University of Colorado. His principal areas of teaching and research are operations research and engineering economics. Dr. Toole holds a BA in mathematics and a PhD in industrial engineering and operations research from the University of California at Berkeley.

CHAPTER 6
COST CONTROL USING ESTIMATES

What will you pay for energy next year?

To understand energy costs, you must first understand energy rates. Then apply the rates to the big energy users in the plant on a seasonal basis.

PHILLIP F. OSTWALD
University of Colorado, Boulder, CO

With the days of cheap energy rapidly disappearing into history, we are all searching for ways to reduce energy costs. Estimating costs of energies begins with knowledge of the rate structure. Rate structure has evolved using the accepted practices of the utility industry and changes from public power commissions prohibiting unreasonable preferences or advantages to a customer. Rates are, of course, the prices for energy service, while the schedule provides the billing for various load conditions. Table I gives many of the basic definitions and formulas necessary for estimating.

In early years, meters were unavailable to measure loads and energy, and so-called flat rates were used. These were simply prices for the use of certain equipment for a specified time—for example, so much for a certain sited lamp per month. In electric metering, an ampere meter was first used by Edison and service

Table I.
Helpful definitions

Ampere is the measure of an electric current or the rate of flow of electricity. As the rate of flow of water in a pipe might be expressed in gallons per second, the rate of flow of electricity in a circuit is so many units of electricity per second or so many amperes.

Capacity factor—Ratio: $\dfrac{\text{average load}}{\text{rated capacity}}$

This gives the average use of equipment as a percentage of maximum possible.
Coincidence factor—
Ratio: $\dfrac{\text{coincident maximum demand}}{\text{sum of individual maximum demands}}$

This is the inverse of diversity factor.
Demand factor—related to customer's load—
Ratio: $\dfrac{\text{maximum demand}}{\text{total connected load}}$

This gives the simultaneous use made of equipment.
Diversity factor—
Ratio: $\dfrac{\text{sum of individual maximum demands}}{\text{coincident maximum demand}}$

This shows the diversity of use of equipment which results when maximum loads occur at different times.
Duration curve. This curve shows for a selected period, say a year, the total number of hours that each amount of load is carried by a system.

Energy losses = I^2R
or $\qquad\qquad = R(I_p^2 + I_r^2)$
where: I_p = power component of current, I_r = reactive component of current.
Kilovolt-ampere. Certain types of electric circuits and apparatus have a characteristic which causes them to take additional current with no increase in the amount of real power supply. The unit of the apparent electric power is the kilovolt-ampere.
Kilowatt (1000 watts) is the unit of electric power and corresponds in principle to horsepower, which is 0.746 kilowatt (746 watts). Power is the rate of doing work. One horsepower is equivalent to lifting 1 pound 33,000 feet in 1 minute. One kilowatt is 1.34 horsepower. The product of amperes and volts gives watts in a direct current circuit. A kilowatt expresses the size of electric equipment and the load or demand which that equipment imposes on an electric supply system. For example, a 100-watt light creates a load or demand of 100 watts or 0.1 kilowatt.
Kilowatt-hour is the unit of electric energy and is the product of power measured in hours. A piece of equipment rated at 1 kilowatt operated for 1 hour consumes 1 kilowatt-hour. A 100-watt light operated for 20 hours consumes 2 kilowatt-hours.

Load factor—Ratio: $\dfrac{\text{Average load}}{\text{Maximum load}}$

This ratio gives electric equipment as a percentage of maximum possible use of the demand. The factor is calculated as follows, assuming 1000 kilowatt-hours per month and a maximum load of 5 kilowatts:

$$\dfrac{1000 \text{ kWh/mo}}{730 \text{ hr/mo}} = 1.37 \text{ kW}$$

was sold on the ampere hour. Since voltage was not considered, actual energy was not measured. Later the induction type watt-hour meter was developed allowing rates based on kilowatt hour consumption. Ultimately kilowatt demand meters were developed allowing both demand and energy pricing.

A special contract may exist between the firm and the utility if the firm is a large customer of energy. Otherwise, blanket rate schedules, according to district, loads, etc., exist for a specific period, usually a year.

Flat demand rate—This type of rate consists of a price for kilowatt or horsepower for a period of time, say a month or year. This rate is negotiated knowing connected load, so metering is not required. The prices may be blocked or discounts allowed with the total gross bill. Flat demand rates are sometimes used for parking lot or street lighting service.

Example—$6/lamp-month for a mercury vapor (7000 lumens) street light.

Method of calculating bill, assuming an installation of 10 lamps—
10 lamps × $6 = $60 per month.

Straight-line meter rate—This rate is a single price per kilowatt-hour. While it is simple, it does not recognize customer demand costs, nor does it provide lower prices for greater use. This rate might be used for off-peak water heating service. Example: $0.03/kWh.

Method of calculating bill, assuming a monthly use of 600 kilowatt-hours—
600 kWh × 0.03 = $18

Block meter rate—This type specifies certain prices per kilowatt-hour for various kilowatt-hour blocks, the price per kilowatt-hour decreasing for succeeding blocks. The rate in Table II is simple and widely applied to residential and small users. In its basic form it does not recognize the demand element.

Table II. Monthly energy charge, example rate.

Block (kWh)		$/kWh
First	6,000	$0.02789
Next	14,000	0.01829
Next	100,000	0.01509
Next	160,000	0.01499
Next	220,000	0.01316
Over	500,000	0.01267

As an example, assume a monthly use of 42,294 kWh—

First	6,000 kWh × 0.028	$168.00
Next	14,000 kWh × 0.018	252.00
Next	22,294 kWh × 0.015	334.41
		$754.41

$$\frac{1.37 \text{ kW}}{5 \text{ kW maximum load}} = 27\% \text{ load factor}$$

Light source. The standard unit of luminous intensity of a light source is called the candle, and luminous intensity is expressed in candlepower. An incandescent street light requires around 0.6 watt per candlepower, large lamps being more efficient than the small sizes. Sodium and fluorescent require less energy than incandescent.

Loss factor—electric—Ratio: $\frac{\text{average loss}}{\text{maximum loss}}$

For a specific load pattern, the loss factor will bear a certain relationship to a load factor.

Periods of use—In analyzing loads, the following periods in terms of hours are used: week = 168 hours; month = 730 hours; year = 8760 hours.

Power factor—Ratio: $\frac{\text{real power in kW}}{\text{apparent power in kVA}}$

Reactive kilovolt-ampere is the inactive component of apparent electric power. The kilowatt is the active component. The reactive kilovolt-ampere is also termed kilovar, abbreviated kvar.

Utilization factor—related to generating plant—
Ratio: $\frac{\text{maximum demand}}{\text{rated capacity}}$

Volt is the unit of electric force or pressure. As pressure is required to push water through a pipe, force or pressure is necessary to drive electricity through a circuit.

Voltage drop = $I\sqrt{R^2 + X^2}$
or = $I(\cos \phi\, R + \sin \phi\, X)$
where I = total current, R = resistance in ohms, X = reactance in ohms, ϕ = power factor angle.

Gas
1 therm = 10^5 Btu, 1 quad = 10^{15} Btu.
Natural gas averages from 1000 to 1100 Btu per cubic foot at sea level.

Parking lot
A dusk-to-dawn parking lot lighting schedule gives approximately 4000 hours of operation per year.

Refrigeration
A **degree-day** is each degree that the average outside temperature for a day is less than 65 F.
Heat factor is the kilowatt-hours per 1000 cubic feet of house content per degree-day.
Approximately 1 horsepower motor capacity is required per ton of refrigerating capacity.
1 ton of refrigeration or cooling capacity is equal to the amount of cooling produced by melting 1 ton of ice in 24 hours. It is equal to the removal of heat at the rate of 12,000 Btu per hour or 288,000 Btu in 24 hours.

Steam
1 boiler horsepower is the equivalent evaporation of 34.5 pounds of water per hour at 212 F.
1 boiler horsepower is equivalent to 33,479 Btu per hour.

Water heating
1 Btu is the amount of heat required to raise 1 pound of water 1 F. 1 gallon of water weighs 8.33 pounds.
1 kilowatt-hour = 3412 Btu.
1 kilowatt-hour will raise the temperature of 4 gallons of water approximately 100 F.

Water power
Horsepower = 0.114 WHE
where W = cubic feet of water per second, H = head of water in feet, E = over-all efficiency of water system and wheel (say 80%).

Table III. Monthly demand charge, example rate.

First	25 kW of billing demand or less	$85.50
Next	75 kW of billing demand/kW	2.82
Next	200 kW of billing demand/kW	2.68
Next	200 kW of billing demand/kW	2.56
All over	500 kW of billing demand/kW	2.26

Table IV. Yearly energy charge, example rate.

First	100 kWh/yr/hp of billing demand, per kWh	$0.077
Next	200 kWh/yr/hp of billing demand, per kWh	0.048
Next	200 kWh/yr/hp of billing demand, per kWh	0.036

Table V. Example rate for commercial gas user.

First	400 cu ft or less	$2.67
Next	1,600 cu ft, per 100 cu ft	0.151
Next	6,000 cu ft, per 100 cu ft	0.084
Next	12,000 cu ft, per 100 cu ft	0.076
All over	20,000 cu ft, per 100 cu ft	0.068

Table VI. Typical monthly rate for industrial use of gas.

First	10,000 cu ft or less	$15.55
Next	90,000 cu ft, per 1000 cu ft	0.741
Next	400,000 cu ft, per 1000 cu ft	0.478
Next	500,000 cu ft, per 1000 cu ft	0.456

The block meter rate usually includes a minimum monthly charge or a service charge.

Demand rate—This rate consists of separate charges for demand and energy, thus recognizing load factor. One variation is to charge a constant price per kilowatt used, similar to the straight-line method. Another method provides for a number of energy blocks with decreasing prices for succeeding blocks, and in which the sizes of the energy blocks increase with size of load such as in Table III.

Example: Assume a demand load of 247.2 kW, say for an office using lighting, general power, and air conditioning—

First	25 kW	$ 85.50
Next	75 kW	211.50
Next	147.2 kW	394.50
		$691.50

Adding the demand to the energy charge calculated using the block meter rate, the total monthly bill is $754.41 + $691.50 = $1445.91.

Special power rates—As many utilities negotiate a rate with a user, their structure can vary, as shown in Table IV.

This rate might be used for certain industrial power requirements. In agriculture, irrigation power is an example. The billing demand is related to the manufacturer's rating of the largest motor.

Many provisions affect the foregoing schedules, for instance power factor, fuel cost adjustment, billing demand determination, transformation to lower voltage, dump power, off-peak, standby, and availability. Power factor, once more important in the billing structure, is now sometimes treated as a penalty cost to an industry if its power factor is detrimental to system load.

Natural gas rates are similar to the electric rates. A commercial rate may be structured as in Table V.

A commercial gas outdoor area lighting service can be charged on the basis of the number of fixtures, and a charge for each mantle over two. But this simple type of a rate structure is severely limited and a block rate structure is used for industrial service, similar to Table VI. While the price depends upon many features, a lower rate is provided if the energy service is interruptible.

An on-peak demand charge for the maximum daily on-peak gas contracted is supplemental to these rates. (On-peak gas is gas used during a curtailment period).

Fuel cost adjustment—The cost of fuel is a major expense and public service commissions sometimes allow utilities to charge for the variations in cost, either up or down. These rate adjustments can be the linear type, $0.055/100 cu ft for gas and $0.00075/kWh for electric energy. Block rates can be imposed by the utility for fuel adjustment as well. The utility is obliged to satisfy state or reporting laws with these adjustments. Another type of bill adjustment to the firm is for line losses similar to the fuel adjustment, if the energy system involves hydropower and the firm contracts for special rates using hydropower.

Returning now to electric service, the general block rate bases the charges for service on total kilowatt-hours and recovers the demand charge in several block steps. This rate is called a declining block rate because the kilowatt-hour rate decreases with increasing blocks of total energy used. The demand rate bases the charge for electric service on total kilowatt-hours used plus the kilowatt demand determined by a 15-minute (or other length such as 5-, 30-, or 60-minute) period in which the maximum use was recorded during the monthly billing period.

Several other demand notions enter into the billing calculation. Load peaks may be for different time intervals. A single peak may be used or several peaks averaged, and off-peak loads may be related to loads created during on-peak hours and taken at reduced value. Many schedules contain features providing for lower charges during off-peak periods to induce customers to divert loads away from the system peak load. A minimum monthly charge is customary.

Example

Consider the problem of estimating a 21,000-square-foot office building having a total connected lighting load of 61.6 kW, general power of 14.1 kW (demand), and a roof-top air conditioner with a total load of 171.5 kW. 30% of this load is for

Lighting and general power	75.7 kW		13,653 kWh/mo
A/C—May, June and Sept.	171.5		28,641
	247.2 kW		42,294 kWh/mo
Demand charge:	100.0 kW	@ $2.68	$297.00
	147.2		394.50
	247.2 kW		$691.50
Energy charge:	20,000 kWh	@ $.015	$420.00
	22,294		334.41
	42,294 kWh		$754.41

Figure 1. Calculating one month electric bill.

heating and ventilating. Office hours are 8 am to 5 pm, 5 days per week.

For lighting, assume an office use of 10 hours per day to include cleaning— (61.6 kW) (10 hr/da) (5 da/wk) (52)/12 = 13,347 kWh/mo. The lighting operates 217 hours per month.

For general power, assume 14.1 kW use, one hour per day, 5 days per week, or 306 kWh/mo, operating 22 hours per month.

For air conditioning (A/C), assume 1000 hours per season, 5 months, May through September. Using 16.7% of load in May, June, and September: 167 hr × 171.5 kW = 28,641 kWh/mo. Using 25.0% of load in July and August, 250 hr × 171.5 kW = 42,875 kWh/mo.

For heating and ventilating, assume 1500 hours per season, 7 months, October through April. Using 12% of load in October, November, and April, 12% (1500 hr) = 180 hr/mo. 180 hr × 30% of 171.5 or 51.45 kW = 9261 kWh/mo. 16% used in December, January, February, and March: 16% (1500 hr) = 240 hr/mo. 240 hr/mo × 51.45 kW = 12,348 kWh/mo. Total of lighting and general power = 13,347 + 306 = 13,653 kWh/mo. Total demand is 75.7 kW.

The above figures are added to the respective months of A/C or heating kW and kWh.

Because of seasonal variations, we end up with four different bills similar to that shown in Figure 1.

Thus: A/C May, June, and September, 247.2 kW demand, 42,294 kWh;

$1,445.91 × 3 mo = $4,337.73.
A/C July and August, 247.2 kW demand, 56,528 kWh;
$1,659.42 × 2 mo = $3,318.84.
Heating and ventilating October, November, and April, 127.15 kW demand, 22,914 kWh;
$833.47 × 3 mo = $2,500.41.
Heating and ventilating December, January, February, March, 127.15 kW demand, 26,001 kWh;
$879.78 × 4 mo = $ 3,519.12.
Total = $13,676.10.

Cost $/sq ft = $\frac{\$13,676.10}{21,000}$ = $0.65.

Manufacturing

Electric loads in the manufacturing plant can be evaluated following these rules of thumb and after gathering information on the load rating. The plant load is broken down into building lighting, air conditioning, and ventilating systems, and manufacturing process loads. The building kW demand = connected lighting kW + air conditioning kW + ventilation kW. In manufacturing, use the connected kW multiplied by the factors which account for simultaneous use and diversity, and add the total of the processes and machines. For heavy users of electric energy, such as arc furnaces or inductor furnaces, motor-generator welders, etc., use rated kW × 1.0. In automated machine shops, kW = connected hp × 0.3. For manual operated machines, kW = connected hp × 0.2. For manual operated machines where there is one operator for several machines, kW = 0.5 × hp of the largest machine for each operator. The demand may arise from large machines or installed electric process heating. In electric process heating, sum the total kW connected, find the kW to start and maintain the process, and determine the frequency of starting. For automatic spot welders, kW = 0.3 × kVA, and spot welding, kW = 0.16 × kVA, where there is one operator per welder.

In the case of an energy saving program, the values of kW demand and kWh energy calculated using these rules of thumb are evaluated using the top blocks in which the demand and energy fall. Assume a monthly bill of $1,445.91 during May, June, and September. If a cost reduction program envisions a persistent demand reduction of 25.2 kW and energy of 7,206 kWh, then savings are:

25.2 kW × $2.68/kW = $67.54
7,206 kWh × $0.015/kWh = 108.09
$175.63

and a yearly saving, if these demands and energy usage are level-month reductions, of $2107.56. These savings do not consider capital, labor, and other expenses that might be offsetting.

Phillip F. Ostwald is Professor of Mechanical and Industrial Engineering in the Mechanical Engineering Dept at the University of Colorado where he has been active in teaching and research, primarily in cost estimating. Also he regularly teaches short courses in manufacturing cost estimating for the American Management Associations. Dr Ostwald holds a BSME from the University of Nebraska, MSIE from Ohio State University, and a PhD from Oklahoma State University. He is a Senior Member of AIIE.

H. K. Von Kaas

Principal

H. K. Von Kaas and Associates

811 E. Wisconsin Avenue

Milwaukee, Wisconsin 53202

PROCEDURES FOR CONTROLLING MANUFACTURING COSTS IN THE DURABLE GOODS INDUSTRIES

When we speak of control of operations we are talking about the process of comparing performance with some type of standard, and developing a means of bringing variances back into line. In a manufacturing environment the factory operating statement usually serves as the display which is the basis for any required corrective action.

Conventional cost accounting procedures are commonly based on preparation of product costs, profit and loss statements, and balance sheets. They are usually more tax oriented than control oriented. For example, the customary factory operating statement merely summarizes money spent. Labor items are frequently totaled under two main headings - direct and indirect. On the other hand, other expense categories are frequently broken up into small subdivisions. The result is shown in Exhibit 1 which is taken from samples in my files. The figures shown are representative of many medium sized durable goods manufacturing situations.

Note that this example has two serious deficiencies. The first is that the two major categories, direct labor and indirect labor, which account for 80 percent of the total, are not broken down in any detail; second, there is no measure of relative output to justify these expenditures. It merely records past history, and is useless as a control. We have no way of knowing whether our factory manager has done a good job or a poor one.

Since the labor accounts contribute to the major expenditures in most manufacturing operations, as shown in the example, good control procedures indicate a breakdown of these accounts into controllable categories. Let us defer our analysis of direct labor for a moment and consider indirect labor. Indirect labor serves two main purposes. It supports the direct labor effort in such areas as material handling, tool maintenance, inspection, factory clerical, etc. It also summarizes excess costs such as re-work, waiting for material, lost time, etc.

This breakdown of indirect labor initiates the beginning of controls for operating departments. These are, of course, a few functions which are completely indirect, such as shipping

EXHIBIT 1

CONSOLIDATED COMPRESSOR, INC.

Factory Operating Statement - Month of _____

Direct Labor	$ 79,281
Indirect Labor	32,747
Payroll Taxes	2,183
Supplies	9,025
General Insurance	2,007
Group Insurance	911
Contract Labor	0
Rent	6,462
Telephone	1,968
Travel	629
Entertainment	93
Dues, Subscriptions	150
Equipment Rental	376
Heat, Light, Water	713
Depreciation	2,400
Sundry	30
Employee Training	25
Janitor	1,332
	$ 140,332

and receiving, also plant maintenance. They require a slightly different approach to the control problem.

Practically all of these categories of indirect labor with the possible exception of maintenance, and a few other exceptions such as janitors, watchmen, etc., are production related. They will generally vary in some ratio to direct labor.

Direct labor is, of course, completely production related. Various ratios can be developed. The simplest ratio is direct labor dollars per dollar of sales. This ratio works well provided there are no major fluctuations in process or finished goods inventory, no significant changes in the wage/price relationship, and if only a single product, or product line with a uniform labor content is involved. It is less reliable where there are changes in product mix, or in a job shop situation where actual direct labor, for example, is compared to estimated labor for an order. Here a second variable, accuracy of estimating, must be considered.

The introduction of work measurement, and incentives, greatly simplifies the control and evaluation of direct labor, particularly when a standard hour plan is used. Standard hours produced then become a common denominator for accurately evaluating direct labor performance not only plant wide, but between departments or divisions. "Standard hours produced" eliminates the variables of product mix, inventory fluctuations, and all the vagaries of wage versus price relationships. In many situations, the development of work measurement and direct labor standards is highly worthwhile even if not reinforced by incentives.

However, the introduction of work measurement and incentives alone does not completely solve the control problem. If only one cost area is controlled, excess labor time can easily be diverted to less controlled areas. Anyone familiar with wage incentive administration has observed instances where an operation, after completing an incentive operation, may be clocked out on material handling or clean up because there is no other incentive work immediately available.

Thus, the introduction of incentives, which result in increased production, generally has two effects. Either output is increased for the same direct labor input, or labor hours (and costs) decrease for the same level of production. This simple fact is frequently overlooked at the foreman level.

Two simple ratios can be used to indicate control, or the lack of control, in this area. We have previously mentioned the categories of support labor and excess labor. The Factory Support Ratio (total support hours ÷ standard hours produced) indicates the ratio of support hours to sustain the productive effort of direct labor. Any diversion of direct labor to support labor magnifies this ratio markedly, since the numerator is increased and the denominator is decreased at the same time. Exactly the same is true of the Factory Excess Ratio.

Factory Support consists of such indirect labor categories as clean-up, material handling, set-ups, tool grinding, inspection, etc.

Factory Excess consists of indirect labor for salvage, rework, lost time for various causes, and items of similar nature.

The formulas are as follows:

   a. Factory Support Ratio = $\dfrac{\text{Factory Support Hours}}{\text{Standard Hours Produced}}$

   b. Factory Excess Ratio = $\dfrac{\text{Factory Excess Hours}}{\text{Standard Hours Produced}}$

Both of these ratios are positive downward, that is, any improvement shows up as a decrease.

The following case study (Exhibit 2) illustrates a typical situation where the actual figures are representative for the purpose of illustration.

EXHIBIT 2

Effect of Incentive Installation

CONSOLIDATED COMPRESSOR, INC.

	A Before Incentives	B After Incentives	
		Condition 1	Condition 2
Plant Hours:			
1. Direct Labor Hours	6,000	4,000	4,000
2. Earned Hours on Incentive	---	4,800	4,800
3. Support Hours—(Set-up-Mat'l. Hdlg. Tool Grinding, etc.)	2,500	3,000	2,500
4. Excess Hours—(Repairs-Rework-Lost Time, etc.)	1,500	3,000	1,500
Total Labor Hours:			
5. Actual (1+3+4)	10,000	10,000	10,000
6. Hours Paid For (2+3+4)	10,000	10,800	8,800
Control Information:			
7. Total Compressors Produced	2,000	2,000	2,000
8. Total Paid Hours per compressor (6 ÷ 7)	5.0	5.4	4.4
9. Total Payroll @ $3.00/hr	$30,000	$32,400	$26,400
10. Total Labor Cost per compressor @ $3.00/hr.	15.00	16.20	13.20
11. Incentive Gains (2 ÷ 1)	---	120%	120%
12. Incentive Coverage (1 ÷ 5)	---	40%	40%
13. Factory Support Ratio (3 ÷ 1)	.42	.75	.63
14. Factory Excess Ratio (4 ÷ 1)	.25	.75	.38

Consolidated Compressors, Inc. employed 60 hourly employees who worked a total of 10,000 hours per month, at an average wage of $3.00 per hour before the institution of incentives. Production schedules were fairly uniform at 10,000 units per month. In an effort to reduce costs, the company installed a standard hour incentive plan which resulted in a substantial cost reduction on all direct labor operations.

Column A shows the situation before installation of the incentive plan. After installation, Condition 1 illustrates what happens in many situations where there is a lack of good control. Here the foreman, who had been supervising this department for many years, felt that his experience qualified him to judge how many people he needed to operate his department for a given schedule.

He had little faith in incentives and was reluctant to reduce his work force. His superiors in management took little interest in effective labor controls and very little time was spent in explaining the working of the incentive system. After the installation of incentives, the production workers began to increase their output and soon achieved satisfactory incentive earnings. However, as a result of increases in their productivity, within a short time they were ahead of schedule.

Not realizing that the parts schedules were based on a relatively constant schedule of 2,000 compressors a month, the foreman began to try to find other work for his men while waiting for materials. Depending on the monetary situation, he would put his men on inspection, rework, material handling, or simply tell them to punch out on waiting for materials.

Because of this, the installation of incentives actually increased the company's total labor cost per compressor from $15.00 to $16.20. The two indicators, Factory Support Ratio, and Factory Excess Ratio, tell the story of what happened. Note that the Factory Support Ratio increased from .42 hours of support labor per hour of direct labor to .75 hours of support labor per hour of direct labor after incentives. Similarly, the Factory Excess Ratio jumped from .25 to .75.

These two ratios are very good indicators in controlling indirect labor. Any substantial increase points out poor labor controls, particularly when surplus labor is diverted to indirect accounts.

In a situation such as the above, it is easy to see why management might conclude that the incentive system was a failure and actually increased costs.

Under Condition 2, we are assuming a situation where the installation of the incentive plan was accompanied by good labor controls and good management techniques. Top management carefully reviewed the operating statements with the foreman. It was pointed out that his department had been producing about 2,000 compressors per month with about 2,500 hours of Support Labor, and 1,500 hours of "Excess" labor and he was encouraged to maintain these levels of indirect labor.

Direct Hours required for production of 2,000 compressors per month would be reduced with incentives, and plans were made to diminish manpower accordingly. This was done by attrition, transfers and weeding out probationary employees. The result was the elimination of 2,000 manhours in the department.

Thus in Condition 2, we have a situation where installation of incentives, plus education and careful manpower planning reduced the total hours per compressor from 5.0 to 4.4, a reduction in labor cost from $15.00 to $13.20 and savings of about $3,600 per month. Under such conditions we can expect management to be well satisfied with incentives, and extend their use to other areas.

The somewhat elementary analysis just described can be expanded to form a new complete Factory Operating Statement shown in Exhibit 3. This permits a wide variety of controls, and is adaptable to nearly all types of durable goods manufacturing. It presumes the use of the typical standard hour incentive plan. It breaks down direct labor by incentive and non-incentive, and includes typical accounts for the various activities included in Factory Support, and Factory Excess Labor.

EXHIBIT 3

FACTORY OPERATING STATEMENT
Month of _____

	Code	Dollars	Code	Hours
Total Hourly Payroll	1	$34,335	24	11,274
Total Clocked Payroll	2		25	10,720
Direct Labor				
Produced on Standard	3	$10,200	26	3,396
Actual Hours on Incentive	4		27	2,842
Actual Hours-Non Incentive	5	11,160	28	3,729
Total Direct Labor	6	$21,360	29	6,571
Indirect Labor				
Factory Support:				
Tools & Fixtures	7	$ 1,274	30	392
Set-Ups	8	607	31	201
Planned Maintenance	9	998	32	322
Material Handling	10	2,620	33	936
Hourly Supervision	11	2,184	34	560
Factory Clerical	12	440	35	160
Total Factory Support	13	$ 8,123	36	2,571
Factory Excess				
Lost Time	14	$ 3,248	37	1,065
Equipment Repairs	15	485	38	152
Material Trouble	16	127	39	41
Salvage & Rework	17	992	40	320
Total Factory Excess	18	$ 4,852	41	1,478
Total Indirect Labor	19	$12,975	42	4,149

Labor Variance (Included in above totals)

	Code	Dollars	Code	Hours
Incentive Make Up	20	461	43	159
Incentive Gain	21	1,689	44	554(26-27)

Incentive make up shows dollar costs, and hours, required to pay minimum (day rate) earnings to sub-standard performers. Incentive gains shows dollars and hours earned in excess of daywork under the incentive system.

NOTATIONS:

Code 1 - Total Hourly Payroll, $,(6+19=1): Hours (29+42=25)
Code 6 - Total Direct Payroll, $,(3+ 5=6): Hours (27+28=29)
Code 19 - Total Indirect Payroll, $, (13+18=19):Hours (36+41=42)

Labor controls which can be developed from this operating statement are as follows:

a. Incentive Gain; 26/27 (3396 ÷ 2842) = 119%; individual jobs should show normal distribution around median gain.
b. Incentive coverage, direct 27/29 (2842 ÷ 6571) = 43% positive upward.
c. Incentive coverage, plant 27/25 (2842 ÷ 10,720) = 26.5%.
d. Factory Overall Production Index = Standard Hours Produced Code 26 (3396).
e. Total Labor Cost per Standard Hour (1/26); $34,335 ÷ 3396 - $10.11.
f. Indirect Ratio 42/26 (4149 ÷ 3396) = 1.22 (it requires 1.22 hours of indirect labor per standard hour produced).

211

g. Factory Support Ratio 36 - 26
   (2571 ÷ 3396) = .76*
h. Factory Excess Ratio 41 - 26
   (1578 ÷ 3396) = .46*
i. Average Hourly Earnings 6 - 29
   (21,360 ÷ 6571) = $3.25.
j. Average Hourly Earnings, (plant)
   1 ÷ 25 = $34,335 ÷ 10,720 = $3.20.

* g and h, Factory Support and Factory Excess ratios were discussed previously and are major ratios for good labor control.

These controls are based on the premise that the standard hours produced, as determined from the route sheets, are the common denominator for measuring factory labor output; and are an excellent measure of the value added by the manufacturing process, regardless of the particular part or product. With minor exceptions, such as factory clerical and hourly supervision and inspection, all other labor accounts are closely related to production levels so that standard hours produced are a common basis for evaluating performance of all other labor accounts.

With the addition of non-labor accounts this statement can then furnish additional manufacturing cost controls, as well as furnish basic data, for use in the Cost of Sales Analysis.

Assume the following as representative of typical non-labor factory overhead accounts:

Factory Overhead Analysis
    (Non-Labor Accounts)

Salaried Supervision	$ 3,000
Payroll Taxes	478
Supplies	1,473
Group Insurance	674
Depreciation	5,210
Heat, Light Power	493
Occupancy	972
Vacation Accurals	1,063
Code 45 TOTAL	$13,363

MANUFACTURING COST CONTROLS: (ALL POSITIVE DOWNWARD)

a. Manufacturing overhead (non-labor) ratio: Manufacturing overhead ÷ standard hours = (45 ÷ 26) = $13,363 ÷ 3396 = $3.93/standard hour.
b. Total overhead cost ratio: Indirect labor + manufacturing non-labor overhead ÷ standard hours
   $\frac{19 + 45}{26} = \frac{\$12,975 + \$13,363}{3396}$ = $7.76 per standard hour.
c. Total manufacturing cost per standard hour produced: $\frac{1 + 45}{26}$ =
   $\frac{\text{total labor + mfg. overhead (non-labor)}}{\text{standard hours produced}}$
   = $\frac{\$34,335 + \$13,363}{3396}$ = $14.04

This latter ratio is particularly useful in developing estimated costs for new or revised products.

This information is now ready to be entered into the Cost of Sales Analysis.

COST OF SALES ANALYSIS

Work in Process - Beginning		$ 59,641	
Materials Used -			
Raw Materials, Beginning	$116,602		
Raw Materials & Components Purchasing	34,983		
Available	$151,585		
Used During Period	68,502	68,502	1
Raw Material, Ending	$ 83,083		
Direct Labor (Code 1)		$ 34,335	
Indirect Labor (Code 19)		12,975	
Contract Work (Plating, etc.)		8,200	
Manufacturing Overhead (Non-Labor Code 45)		13,363	
Total to Work in Process		$197,016	
Less Work in Process Ending		106,857	2
Finished Goods Manufactured		$ 90,159	
Finished Goods Inventory, Beginning		83,718	
Available for Shipment		$173,877	
Less Finished Goods Inventory, Ending		45,706	3
Cost of Goods Sold		$128,171	

a. Summary of all materials at Standard Cost, from Route Sheets for parts manufactured (Summary "F" on Route Sheets).
b. Beginning Work in Process inventory plus additions, minus products transferred to Finished Goods ("L" on Route Sheet).
c. Finished Goods Beginning plus transfers to Finished Goods ("L" on Route Sheet) minus shipments at Standard Cost.

MANUFACTURING PROFITABILITY

Manufacturing profitability is defined as the margin between net sales and cost of sales. Manufacturing profits must be sufficient to cover all general and administrative costs plus necessary corporate profits. Msnufacturing profitability forms an excellent yardstick for supervisory incentive payments.

The usual calculation:

Net Sales	$217,000	100%
Cost of Sales	128,171	59%
Manufacturing Profit	$ 88,829	41%

Procedures for installing controls of this nature require two important preliminary steps.

The first step is the development of a Chart of Accounts for the manufacturing operations which will properly segregate the various direct and indirect labor accounts. Exhibits 4 and 5 are typical of a Chart of Accounts for these labor categories for the usual durable goods manufacturing enterprise. Note that the direct labor accounts are segregated by major departments as well as by standard and non-standard labor hours. Similarly, the indirect labor accounts are broken down by various functions under the general headings of Factory Support and Factory Excess. Actual account numbers and specific labor accounts will, of course, very between different industrial firms.

EXHIBIT 4

TYPICAL CHART OF ACCOUNTS
DIRECT LABOR

New Account	
5010	Machine Shop (In plant labor) on standard.
5020	Polishing & Plating (In plant labor) on standard.
5030	Assembling (In plant labor) on standard.
5035	Customer repair service (In plant labor) on standard.
5040	Final Assembly new models (debugging).
5045	Salvage Labor (In plant labor).
5050	Machine Shop off standard (new jobs, pilot production runs, added operations, mechanical/material problems).
5055	Machine Shop D/L excess (makeup or daywork).
5060	Polishing & Plating off standard (new jobs, pilot production runs, added operations, mechanical/material problems).
5065	Polish & Plate excess (makeup or daywork).
5070	Assembling off standard (new jobs, pilot production runs, added operations, mechanical/material problems).
5075	Assembling excess (makeup or daywork).

EXHIBIT 5

TYPICAL CHART OF ACCOUNTS
INDIRECT LABOR

Factory Support

Account No.	
6100	Set-up
6105	Material Handling, Moving Stock (In plant).
6110	Stock Room.
6115	Tool Room.
6120	Machine Repair and Tool Repair.
6125	General Maintenance and Millright.
6130	Tool and Cutter Grinding.
6135	Watchman, Janitor - Plant.
6140	Truck driver, fork lift operator.
6145	Engineering and Experimental (samples, tryout, test).
6150	Inspection.

Factory Excess

6160	Rework (defective workmanship).
6165	Rework (defective vendor material).
6170	Non-productive (clean-up, washing, inventory).
6175	Downtime (material, machinery or tool delays).
6180	Lost time (first aid, meetings, paid absence).

The second step in implementation lies in the requirement of accurate shop recording. Shop records and controls are useless unless they define accurately what work was done, starting and stopping times, as well as strict accounting for quantities of completed work. Unless this is done in such a manner that the information is subject to audit and verification, the system is open to abuse and misuse, particularly when incentives are being used.

In summary, proper cost control in the durable goods industries involves three significant factors:

a. Proper segregation of the various cost factors into recognizable and controllable categories.
b. Relating these input costs to some sound measure of output.
c. Insuring the proper recording of labor charges and completed work at the factory floor level by procedures that are subject to verification and audit.

Reprinted from Management Accounting, January, 1976.

LIFE CYCLE COSTING

In its broadest application, an LCC analysis may be used to estimate all relevant costs, both present and future, in the decision-making process of selecting from among various alternative products.

By Gary E. White and Phillip F. Ostwald

The life cycle cost (LCC) of an item is the sum of all funds expended in support of the item from its conception and fabrication, through its operation to the end of its useful life.[1] In its broadest application an LCC analysis may be used to estimate all relevant costs, both present and future, in the decision-making process of selecting from among various alternative products. For example, this analysis would be performed by a manufacturer who is contemplating development and production of a new product for use or general sale, or considering different design characteristics for special-order products. The benefit of an LCC analysis lies in the framework it provides for specifying the estimated total incremental costs of developing, producing, and using a particular item. The life cycle costs thus provide the "pool" of relevant cost information for particular decisions related to the item such as pricing, make-or-buy, capital investment, and many others.

Life Cycle Cost Structure

Usually, the life cycle cost structure is separated into three major components. These consist of engineering and development, production and implementation, and operation. The detailed costs for each component will depend on the particular project, system, or product under consideration. The key principle, in any case, is that these costs represent the estimated direct and incremental costs associated with the item. Exhibit 1 illustrates the three major cost components of life cycle costs as related to time. Certain underlying assumptions concerning the rate and timing of the various costs which necessarily vary among different items are evident. The total estimated future costs at any point in the life cycle would be based on discounted present value estimates which could be used for comparison with other decision alternatives. For example, operating funds necessary to implement a program can be greater than the original research and development investment money required. Moreover, a product with higher development and investment costs but lower operating costs may, depending on service life, be a least-cost product. An LCC analysis also encourages a trade-off between one-time costs and recurring costs.

Example

Assume that an aerospace firm has issued a request for bids on a vacuum chamber design to simulate high altitude pressures for electronic equipment. Two firms have submitted conceptual designs and LCC analyses. The aerospace firm, having more than one technically qualified vendor is able to perform an LCC trade-off study to ascertain the vacuum chamber with the lowest LCC.[2] In preparation for the design bid the following data are required for an LCC analysis.

1. Cost elements
2. Operating profile of the user
3. Utilization factors
4. Costs at current prices
5. Current cost projections with inflation indices
6. Costs discounted to base period, and discount factors
7. Sums of discounted and undiscounted costs

The vendors estimate and classify the cost elements into the three major categories of engineering, production, and operation. We will not concern ourselves with the vendors' cost estimating problems, but rather we assume that the vendors have provided bid prices which include their share of the costs, primarily engineering and production, plus a profit. In addition to the prices, the vendors provide the data contained in Exhibit 2.

Note that some of the factors indicate operating costs incurred during the life of the equipment. Two

---

[1] The concept of life cycle costing has been employed by the Department of Defense in analyzing the cost effectiveness of various systems and projects and its principles can also be used in commercial applications.
[2] It is necessary that the vendors and the firm develop a mutual trust in creating these data. Cooperation is required, and the firm must protect the vendor's cost data by ethical standards of information conduct.

G. E. WHITE

is an Associate Professor of Accounting at the College of Business and Administration, University of Colorado, Boulder, Colorado. He holds a Ph.D. degree in Accounting from the University of Washington.

P. F. OSTWALD

is a Professor of Engineering Design and Economic Evaluation at the College of Engineering and Applied Science, University of Colorado, Boulder, Colorado. He holds a Ph.D. degree from Oklahoma State University.

This article was submitted through the Denver Chapter.

common cost parameters are mean time between failure (MTBF) and mean time to repair (MTTR). Power consumption rate and preventive maintenance routines such as the cycle and the preventive maintenance rate are additional required information.

The firm provides the operating profile for the equipment. This permits calculation of the periodic maintenance experiences. The operating profile has a repetition time and contains all the operating and nonoperating modes of the equipment. It is sometimes possible to have operating profiles internal to other operating profiles. For trade-off studies, vendors must be evaluated with the same operating profile. In our case, this means that we must know our usage of the vacuum chamber. The operating profile tells when the equipment will be operated; the utilization factors tell us how or in what way the equipment will be operating during each mode. For example, in the vacuum chamber, the diffusion pump may run 60 percent of the time; while in the shutdown mode it may not run. Instead, a roughing pump may hold a nominal vacuum.

The aerospace firm estimates its own operating profile, and determines that a continuous three shift operation for about one-third of a year of 2,920 hours per year for two years is necessary. This is applicable to both vacuum pumps. Using the manning requirements and labor rates (accounting for time and a half and fringe costs) it finds that vendor A's product requires 2,920 hours per year times eight dollars per hour times one man per machine equals $23,360. Vendor B's product requires 2,920 hours per year times eight dollars per hour times two men per machine equals $46,720.

Labor for preventive maintenance (PM) actions is calculated from the formula

$$\text{Number of PM actions} = \frac{\text{Scheduled operating hours}}{\text{PM cycle time}}$$

For vendor A's product the number of preventive maintenance actions is 2,920/160 equals 19. Since each action uses four hours, we would have 76 hours of PM time for a total yearly cost of $456 (76 hours times $6 per hour). For vendor B's product the yearly cost is $816.

The vendors supply information indicating that vendor A's product will fail every 500 hours and vendor B's product will fail every 300 hours. These figures are the mean time between failures. The cost of corrective maintenance can be found using the formula:

$$C_{CM} = \frac{SOH}{MTBF}(MTTR)C_{me}$$

where

C_{CM} = Cost for corrective maintenance/year

SOH = Scheduled operating hours
$MTBF$ = Mean hours between failures
$MTTR$ = Mean hours to repair
C_{me} = Cost of maintenance labor

For vendor A's product we have:

$$C_{CM} = \frac{2{,}920 \text{ hrs}}{500 \text{ hrs}}(40 \text{ hrs})(\$6/\text{hr}) = \$1{,}400$$

For vendor B's product the annual costs are computed as $4,672.

Inasmuch as a two-year period is concerned, proper economic analysis makes it imperative that the future estimated costs be discounted to present values. (We have omitted this portion of the analysis due to its wide coverage in other sources.) Similarly, the future effects of inflation or deflation on recurring costs should be considered and then discounted back to the present time. Exhibit 3 presents the extended data for a comparison of the

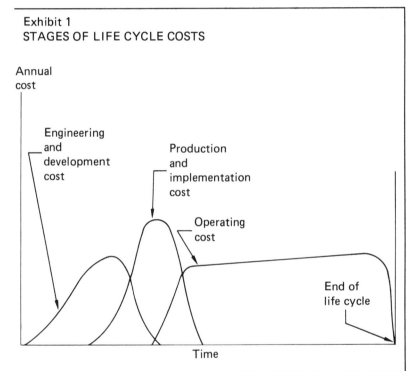

Exhibit 1
STAGES OF LIFE CYCLE COSTS

Exhibit 2
COST DATA FOR LCC ANALYSIS

Cost item	Data source	Estimate vendor A	Estimate vendor B
Product price	Mfgr	$200,000	$170,000
Equipment life	Customer's specification	2 yrs	2 yrs
Installation	Owner	$3,000	$4,000
Manning	Mfgr	1 man	2 men
Manning labor rate	Owner	$8/hr	$8/hr
Mean time between failures	Mfgr	500 hrs	300 hrs
Mean time to repair	Mfgr	1 wk	2 wks
Preventive maintenance cycle	Mfgr	160 hrs	180 hrs
Preventive maintenance downtime	Mfgr	4 hrs	8 hrs
Maintenance labor rate	Owner	$6/hr	$6/hr
Parts and supplies cost (% of product price)	Mfgr	1%	2%
Input power	Mfgr	8.0 kw	9.0 kw
Cost per kwh	Owner	$.025	$.025

LIFE CYCLE COSTING

Continued from page 40

products of the two vendors. Only the incremental costs which are estimated to differ between the models should be included in the analysis. The higher initial cost of vendor A's product with its better maintenance performance and lower operating cost returns the economy in terms of operating performance and has the lowest LCC.

Conclusion

The chief benefit of an LCC analysis consists of the overall framework for considering total incremental costs to be incurred during the life of a project, product or system, and the focus that the analysis places on operating parameters, e.g., operating profiles, utilization factors, mean time between failures, mean time to repair, and maintenance cycles. Thus, LCC emphasizes very careful estimation of operating costs in addition to the initial investment in development and production costs. □

Exhibit 3

LCC TRADE-OFF ANALYSIS OF TWO COMPETITIVE PRODUCTS FOR TWO YEARS

Cost	Vendor A	Vendor B
Product price	$200,000	$170,000
Installation	3,000	4,000
Manning labor (two years)	46,720	93,440
Preventive maintenance (two years)	912	1,632
Corrective maintenance (two years)	2,800	9,344
Power requirements (@ .025/kwh.)	1,168	1,314
Parts and supplies cost (@ 1% and 2% of product price respectively)	2,000	3,400
Total	$256,600	$283,130

Reprinted from Manufacturing Engineering, May, 1977.

Your Cost System: A Management or Accounting Tool?

When properly designed and used, a standard cost system can be of substantial assistance in the areas of cost management and decision making. Here's how they're structured — and how they're used

ROBERT L. DEWELT
Corporate Manager
Cost Systems & Inventory Control
Allis-Chalmers

The NEED FOR A COST SYSTEM to provide financial information for income determination, inventory valuation and the preparation of financial statements has been evident to management and accounting personnel for a long time. But in many cases, the techniques employed to meet these needs have centered around the accumulation of actual costs, after they were incurred, for financial statement purposes. This approach met the basic financial requirements of a cost system, but provided little or no information for performance measurement or decision making. As a result, operating management was not provided with information needed for cost control or the measurement of operating performance.

This created a demand for a new type of cost system that would permit the establishment of meaningful benchmarks that could be used to monitor day-to-day performance and signal when deviations from predetermined policies were occurring. Operating management began to recognize that physical and administrative controls became more effective when they were tied to financial controls. This led to the introduction and use of standard cost systems.

STANDARD COST SYSTEMS

It became evident that a cost system could be structured to provide measurements for all significant elements of cost. This permitted management to quantify the financial impact of their various operating decisions while at the same time meeting the basic requirements of financial statement presentation. Through the use of a standard cost system, management is able to report in their financial statements on the same basis as they measure day-to-day operating performance.

The System Defined. A standard cost system is a tool that permits management to determine how effectively the organization is carrying out the predetermined policies and procedures established for its operations. The real effectiveness of this tool is determined by how accurately and timely it informs them of what is *actually* taking place in their operation. If it is not based on the actual techniques employed in operating the business, its measurements will be of little value in monitoring actions and identifying deviations.

With the many different types of businesses in existence today, no single approach to a standard cost system is appropriate in all operations. But because standard cost is a concept rather than a rigid program, it can be molded to meet the specific criteria of any given operation. It can be applied to part or all of the business. It can be limited to stock items, or it can be expanded to encompass all manufactured and purchased parts. It can be used for material costs only or it can include material, labor and burden costs. To be effective, however, it must be oriented to the needs of the business and it must address the problem of the proper presentation and classification of costs in a way that is both meaningful and timely.

Advantages and Objectives. The management in any successful organization must regularly establish objectives for every significant element of their business, and measure the performance of the enterprise in attaining those objectives. A standard cost system can be used as the vehicle to establish objectives for the operation of the production and manufacturing facilities, and to measure performance against these objectives through the determination of variances. It can also be used to measure performance against objectives established for management areas outside of the manufacturing area by reporting changes in operating standards and product costs resulting from management action and decision making. The following is a summary of some of the advantages and objectives of a standard cost system and how they serve management in operating the business, establishing objectives, and measuring performance against these objectives.

1. *The cost system should be integrated to combine the bill of material, labor routings, engineered standards, productivity and costs into a single system of manufacturing control.* The same standards can then be used for performance reporting, standard product costs, inventory valuation, leadtimes and throughout the accounting system.

2. *The cost system should provide measurements for each significant element of cost and permit the evaluation of operations at each level of management.* Standards should be detailed to the extent necessary to permit meaningful accumulation and reporting of costs on a basis consistent with the assignment of responsibility. The same standards should be used for product costs and operating control.

3. *The cost system should provide for the control and reporting of costs by department as incurred.* It can, however, retain the benefits of a job order cost system, since operational costs can be accumulated for each production order and part number, upon completion, to permit analysis.

4. *The cost system should be based on engineered standards established on a consistent basis at attainable levels.* The standards should be detailed to the extent necessary to develop accurate product costs and measure operating efficiency for the wide range of activity in the plant. Variances from standard should be broken down by type and responsibility. Accountability for the elimination of nonstandard conditions must rest with the responsible supervisors and department managers.

5. *The cost system should provide a means of measuring the effect of cost improvement projects on product cost.* This will improve the contribution of the cost improvement program by incorporating product or process cost improvement projects in the current standards, which are then used for measuring operating or manufacturing performance. The cost system then works as a project monitor to determine how well the manufacturing area is able to implement the cost improvement project and operate under the revised conditions. Standards revision variance can then be used to determine the actual savings incurred through cost improvement projects.

6. *The cost system will provide product cost and profitability analysis information.* Standard product costs should be developed for the full range of products manufactured. It can then be used for inventory valuation, costing production, costing sales, making product comparisons, establishing selling prices, and determining the profitability of various products, and product lines. This makes possible a uniform comparison of the cost and profitability of the various products in the product line to promote selective selling.

7. *The cost system will provide a basis for physical inventory valuation.* Standard costs should be developed on an operation basis and be applied to the actual physical quantities. The inventory can then be costed at old and new standards to determine the actual inventory increase or shrinkage, and the effect of standards changes.

8. *The cost system will provide a central file* containing all current orders, shop routings, bills of material, and standards to permit use of this data for the preparation of manpower requirement reports, machine load reports, tooling requirement reports, in-process inventory control reports, and provide the basis for a labor ticket audit system.

BASIC INGREDIENTS

The first step in any attempt to determine cost is to identify what is to be made, the components required, and the functions to be performed. This requires formal documentation of the manufacturing system. The key ingredients of the system are the bill of materials, shop routings, master schedule, departmental budgets, material and labor costs, and burden rates.

Two Standard System. In basic form, the cost should be a two-standard system with a current and base standard. The base standard represents a formal identification and recognition of the manufacturing procedure at a point in time. Base standards should be established once each year at physical inventory time, and remain in effect until the next physical inventory. Standard costs should be developed for each purchased part, raw material item, and repetitively produced manufactured part, assembly and finished product. The base standard represents the predetermined measuring point from which all variances will ultimately be generated. It identifies changing trends in cost levels and becomes the basis for inventory valuation as well as production and sales costing.

The current standard represents the latest cost and manufacturing procedures used for production. It is updated continuously for shop routings and to incorporate the latest changes in the bill of materials. This constant updating is essential if the system is to provide a meaningful measurement of the manufacturing function. To be effective, performance measurement must be made against standards that always reflect current conditions. Periodically, current prices and labor rates can be utilized to provide up-to-date information for product pricing.

MATERIAL COST

Development of a standard cost system

requires a comprehensive bill of material, and an engineering change system, to document the current makeup of the product. Once the bill of material has been established, level codes should be assigned to each component. Since any given component can be used in various products at different levels, it is necessary to determine the lowest level at which a part can be used in any item. This level code is an essential ingredient for using the implosion or single level technique of setting standards. This approach normally proves far superior to the indented bill of material approach when standards are being established for the entire product line.

In the single level approach, costs are imploded from the lowest level up to the highest level. Under the indented bill approach, costs are exploded from the top down. By using the implosion or single level technique, costs are established and reviewed only once at the first point they are used in the system. They can then be used over and over again as necessary without subsequent review.

Purchased material items normally make up a major portion of the cost of all manufactured products. Control over this segment of cost and identification of changing economic trends, on a timely basis, is critical to continued profitable operations.

To aid in the establishment and analysis of material standards, a Purchased Parts History Report should be prepared. This report summarizes all purchases by part number for the inventory year. At the time of setting new material standards, a prepunched card can be created from this report for each part number identifying the base standard cost and the last actual price paid for that part number. An ABC listing can be prepared for all purchased materials, using either forecasted annual usage or prior year actual usage. The top 90% of the dollar value of purchases, which normally accounts for only 20% to 30% of the total part numbers, should be reviewed in detail by the responsible buyer in the Purchasing Department. The remaining items can be reviewed against the Part Number History Report by the Cost Accounting Department to determine if the last purchase was a representative purchase. Where the last purchase cost is found to be representative, it should be used as the new standard.

Use of this technique greatly speeds up the process of setting standards and provides a prepunched input card for all material prices. Only those items requiring a change from the last actual purchase price paid require any special handling. This procedure will permit the rapid development of base standard costs once a year. It will also permit the development of current standard costs at current prices, on a memo basis, at least quarterly. This will help to identify the impact of changing cost levels on product costs for pricing and profitability analysis purposes.

LABOR AND BURDEN COST

The next element needed for maximum effectiveness of the cost system is labor routings containing engineered standards. The use of engineered standards should be an objective because they help to identify what it should cost to manufacture the product under normal conditions with no unusual interference. But the lack of engineered standard hours should not preclude the development of standard costs for labor and burden. Where engineered standard hours are not available, historical actual hours can be used to develop representative standards.

Benefits Without Standards? It is sometimes felt that without engineered standards, the use of standard labor and burden costs will provide no benefits. This is not true, however, because the system will measure whether the operation is becoming more or less efficient than it was in prior periods. In addition, it will measure the extent of any changes in efficiency. It also acts as a monitor of the progress your organization is making in implementing methods improvements and engineered standards for cost reduction and profit improvement purposes. The impact of these standards changes is identified through the methods change variance by relating the current to base standards. Manufacturing department performance relative to these revised standards is measured by the efficiency reports. This permits an evaluation of the adequacy of the standard as well as

the ability of the employee to perform to the standard. Therefore, proper use of a standard cost system can be an invaluable aid in implementing improved Industrial Engineering techniques.

Standards Development. In initially setting standards, many concerns have found it easiest to set standard hours on a total piece part basis. This data is normally readily available and it permits rapid implementation of a basic system. The system can then be expanded to include standards by operation. The operational standards on the labor routings should be stated in hours rather than dollars. This permits use of this data for leadtime calculations in the MRP system and for development of manpower planning data. It also facilitates redevelopment of standard costs as labor rates change, and it permits the calculation of labor rate variances.

Labor rates should be loaded by wage grade, whenever possible, and should reflect the normal rate of pay for an experienced operator. Burden rates should be developed by department, with separate variable and full burden rates. The variable burden rate is necessary to calculate out-of-pocket setup cost and for product contribution and direct cost analysis.

SETUP COST

Setup standards should be established on a per-occurrence basis and be separately identified on the routing. Setup can be allocated back to the product on the basis of the standard lot size. This allocation of setup cost to product cost can be accomplished by using the fixed lot sizes currently employed in the material control system or from the traditional EOQ formula.

The technique, then, is to calculate the per-occurrence setup for the product from the routing, and divide it by the order quantity (normally EOQ or fixed lot size) to determine the average setup cost per piece produced. This is where the master schedule begins to play an important part in determining product costs and setting standards. An explosion of the master schedule provides the forecasted annual usage for the coming year, which is a requirement of the EOQ formula. That dictates that the master schedule include at least one year's requirements of machinery and service parts to operate the system.

For those operations that don't have the ability to explode a master schedule to obtain forecasted gross requirements

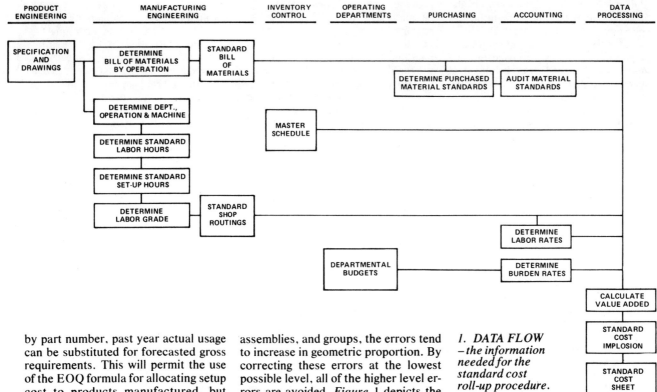

1. DATA FLOW – the information needed for the standard cost roll-up procedure.

by part number, past year actual usage can be substituted for forecasted gross requirements. This will permit the use of the EOQ formula for allocating setup cost to products manufactured, but could result in some distortion of cost on new or obsolete items. Therefore, substantially more analysis of the lot sizes to be used for setup cost allocation will be required to assure representative standard costs.

FILE EDIT

Many people have found it useful to edit certain data in their input files before actually beginning the cost roll-up procedures. Basic problems such as (1) *do all manufactured parts have routings loaded and lot sizing data available?* (2) *are routings incorrectly loaded for any purchased parts?* (3) *do requirements exist for parts that do not currently have a standard and are not included on the Purchase Parts History Report?*, can be detected and corrected prior to attempting a cost roll-up or implosion to eliminate many unnecessary errors later.

STANDARD COST ROLL-UP

Now that all of the input data has been developed, we're ready to begin the standard cost roll-up procedure. This is accomplished by preparing a value added listing which computes the labor and burden cost added at each operation according to the routing. Then, starting at the lowest level, we begin the process of setting standard costs. It is suggested that corrections be made at each level before proceeding to the next level. Since any given part may go into many higher level subassemblies, assemblies, and groups, the errors tend to increase in geometric proportion. By correcting these errors at the lowest possible level, all of the higher level errors are avoided. *Figure* 1 depicts the flow of data necessary for the standard cost roll-up procedure.

USING STANDARD COSTS

After the basic process of setting standards has been completed, you're ready to begin using standard costs in the operation of the business. The premise behind the standard cost concept is management by exception. The standards incorporate the techniques and policies to be employed in the operation of the manufacturing area. Deviations from these predetermined techniques and policies are measured through variances. The reports highlight the very efficient as well as the very inefficient operations and products (the exceptions) as opposed to those operations and products meeting the standard conditions. Management can then concentrate on these exceptions.

Developing the techniques for using standard cost data for day-to-day measurement can be approached in several different ways. The order-closing technique which is such an important element in an actual cost system is also required for an effective standard cost system in many job order environments. Many locations that have encountered problems in converting to and using a standard cost system were often inadequately prepared and gave up a complete actual cost system for parts of a standard cost system.

The conversion to a standard cost system should be approached in modules moving from one phase to another. The initial phase should include a basic order closing routine to remove from inventory all costs that will not otherwise be removed through the cost of sales entry. You cannot take an imprecise environment like a manufacturing area and use systems that presume precision and expect to get the right answer. The order-closing technique removes from inventory the reporting inaccuracies that are inevitable in any manufacturing operation, and permits the use of standard costs for measurement in job order as well as process type environments.

IMPLEMENTATION

By approaching standard costs in a several phase implementation schedule, you can deal with the major problems encountered in each phase independently and accomplish a smooth yet timely conversion. A normal progression would be as follows:

▶ PHASE I. Establish base standard costs to be used in place of estimated actual costs. Identify economic variances such as purchase price variance and labor rate variance as they occur and determine all other variances at the time of order closing.

▶ PHASE II. Develop a technique to maintain current bill of material and

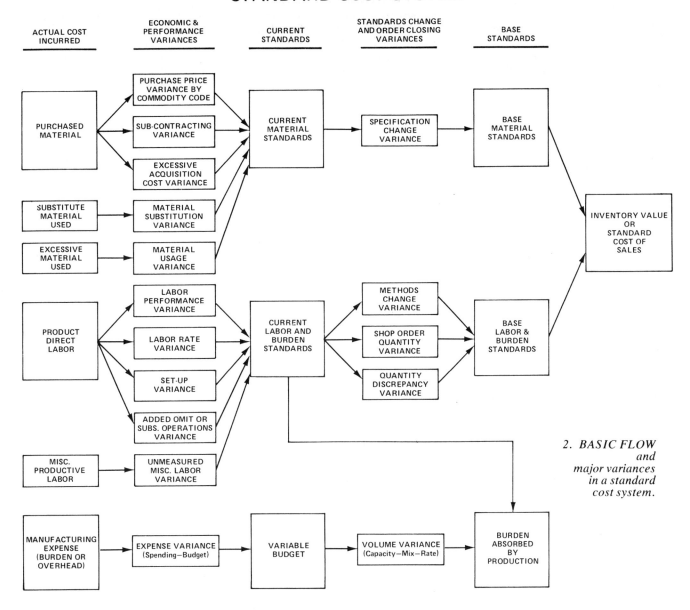

2. *BASIC FLOW and major variances in a standard cost system.*

routing data for improved operational control, pricing decisions, and management information.

▶ PHASE III. Utilize current standards in the labor measurement and control system for improved operational control and the timely calculation of efficiency variances.

▶ PHASE IV. Utilize current standards to develop up-to-date product costs for pricing and product contribution analysis purposes.

In the development of a cost system, it is useful to differentiate between the concept of *cost standards* and *standard costs*. A cost standard is a targeted cost which normally reflects historical actual cost levels. As such, it includes normal inefficiencies encountered in the past, and only measures whether performance is getting better or worse than in prior periods. It often provides an aggregate measurement which does not include the detailed elements necessary for appropriate analysis or determination of the reasons and responsibility for variances.

A standard cost is a predetermined cost that recognizes normal operating conditions and identifies the operations required and time allowed to produce a given product. It is an expansion of the concept of responsibility accounting.

Standard costs are normally developed from standard hours established by using engineered time studies or other acceptable Industrial Engineering techniques, and represent realistic and attainable operating levels. It is necessary however, to differentiate between the concept of realistic and attainable individually or collectively when establishing and using standard costs. The use of standard costs that are realistic and attainable collectively indicates that performance on an overall basis will approximate 100% of the standard cost. To accomplish this, an assumed level of inefficiency must be factored into the standards to compensate for the expected level of performance to the standard.

The use of standard costs that are realistic and individually attainable means that under normal operating conditions each rate is achievable by itself. As such, operating performance to the standard cost, on an overall basis, will be unfavorable due to material shortages, defective tooling, machine breakdowns, new employee training, employee inefficiency and other factors inherent in a normal manufacturing environment. This means that variances must be recognized and budgeted as a separate element of cost.

Use of standards that are realistic and individually attainable permits management by exception for cost system analysis. It permits manufacturing interference and inefficiencies to be identified and quantified as variances for responsibility determination and corrective action where appropriate. Financial statements then report costs on a basis consistent with actual day-to-day operations.

Standard costs become a useful tool for quantifying the financial impact of various situations occurring in the manufacturing environment. Management can then concentrate on the most significant items first, and develop effective strategies for dealing with these items. Using the concept of diminishing returns, management can also determine when the cost to eliminate certain inefficiencies is not justified by the cost of those inefficiencies.

VARIANCE CALCULATION

To make variances (exceptions) truly representative of current conditions and usable for management control purposes, the basis for measurement of floor operations should always be the current routing file. Actual labor tickets can be matched to the current routing file as they are submitted. This permits a measurement of floor operations in terms of what the Industrial Engineering Department has defined as necessary and appropriate to the manufacture of the product. Performance variances are then recognized and recorded on a timely basis. Then, at the time of closing the order, a reconciliation is made from current to base standards. This eliminates the need to do a new cost roll-up of standards every time a routing changes, and permits management to quantify the impact of methods changes, changes in standard hours, or deviations from the planned shop order quantities.

Variances between the actual production cost and the current standard represent operating performance or manufacturing responsibility and should include setup variance, labor performance variance, and added, omit or substitute operations variances. Variances between the current and base standards will be Engineering or Production Control responsibilities and should include items such as specification change variance, methods change variance, and shop order quantity variance. Economic factors such as labor rate variance and purchase price variance are calculated and can be used to determine their financial impact on future product costs and profits. Additional variances such as material substitution variance, material usage variance, volume variance, and expense variance can be calculated and reported for proper responsibility determination. *Figure* 2 shows the basic flow of costs and the major variances to be recognized in a standard cost system.

In some cases, it may be desirable to further breakdown variances into their detailed elements. For example:

▶ Labor performance variance could be segregated into deviations from engineered standards and deviations from estimated or temporary standards. Where a new engineered standards program is being implemented, or where incomplete standards coverage exists, it may be determined that a major portion of the unfavorable variance occurs on jobs that are not covered by permanent engineered standards. Segregation of the variance in this manner may permit a more accurate determination of the underlying cause for the variance. This type of breakdown can be readily obtained by coding the routing to identify engineered versus estimated standards and picking up this identification at the time of calculating the allowed current standard.

▶ Purchase price variance can be segregated by commodity code and/or buyer code to facilitate analysis. This breakdown can be readily obtained by picking up the commodity code and/or buyer code from the item master record.

▶ Special transactions such as subcontracting, or unusual costs such as vendor overtime, can be identified on the purchase order with a code. This code can be picked up at the time of vouchering the vendor's invoice and will permit segregation of these items in the purchase price variance report.

▶ Changes in standards can be identified by reason code at the time of submission, and variances accumulated by these reason codes at a later time. But the cost of identifying reasons for current standard changes, and the cost and practicality of obtaining the detailed actual results must be balanced against the value received from these detailed variances. In most cases, particularly when substantial maintenance or reporting is involved, the additional information may not warrant the cost. It is important to remember, however, that you only get out of a cost system what you put into it. If you want meaningful information for management decision making purposes, you must define the significant elements of cost in your operation and establish measurements and monitoring techniques to provide the appropriate detailed analysis to management on a timely basis.

CURRENT PRODUCT COSTS

The next consideration is providing current cost data to management for pricing and special customer quotes. To accomplish this, a complete roll-up of current standards, on a memo basis, should be attempted at least quarterly. This data can then be used to update your product contribution reports on both a total cost and direct cost basis. If information is required during an interim period for special customer quotes, the indented bill of material can be exploded with major items updated for current changes.

MAXIMUM EFFECTIVENESS

To be effective, the use of standard costs must be a total concept of accounting. It represents an expansion of the basic planning process and tries to explain the reasons for deviations from plan that are occurring on a day-to-day basis. It then permits management to react to these deviations by taking corrective action, or by developing alternate strategies to accomplish their original objectives.

For maximum effectiveness, the system must be dynamic and flexible. It must permit management to change its emphasis as the needs of the business change. One of the advantages of using an integrated system is that instead of each function trying to solve its own problems, the various areas police one another, and must operate within a system where top management has specified the parameters through establishment of rules and policies which are monitored by the computerized systems.

In the highly competitive economy in which we operate today, cost management is a necessity. When properly designed and used, a standard cost system can be of substantial assistance in the areas of cost management and decision making. In the final analysis, the whole purpose of a cost system is to assist in evaluating each manager's effectiveness in controlling those aspects of the total operation for which he is accountable. It bridges the gap between the information needed for accounting control and the needs of operating management. ∎

For maximum effectiveness, the system must be flexible enough to permit management to change its emphasis as the needs of the business change.

Improve your cost estimating by examining and quantifying the uncertainty in each estimate. The cost estimating method described here allows for offsetting estimates and pinpoints areas that contribute the greatest degree of cost uncertainty.

JIM D. BURCH, Honeywell Information Systems, Phoenix, Arizona

Reprinted with permission from Industrial Engineering Magazine, (March 1975). Copyright American Institute of Industrial Engineers, Inc., 25 Technology Park/Atlanta, Norcross, GA 30092.

Cost estimating with uncertainty?

If the following story sounds familiar, perhaps you will be interested in the technique described for cost estimating with uncertainty.

The estimated direct material cost for new product XYZ is $459.87, and estimated direct labor is $1,000. Total direct cost is therefore estimated to be $1,459.87. After production of XYZ starts, we find that the direct material cost is $459.87 but direct labor is $1,550, for a total direct cost of $2,009.87. Upon investigation, we learn that the material cost estimate was prepared by applying standard costs to the parts list, a process requiring eight man-hours, and the labor estimate was obtained by looking at the ceiling in the planning office for ten minutes. Before we fire the labor estimator, we learn that, although complete purchased parts lists were available at the time the estimate was required, no assembly drawings had yet been made, and the accuracy of his estimate would not have improved if he had looked at his ceiling all day. Maybe our displeasure should be directed toward the person responsible for expending eight hours to determine with great precision the direct material content, whose contribution to the total is less than the total cost uncertainty.

If you always have complete product documentation for cost estimates, and if you are comfortable with estimates (or "actuals" for that matter) that imply six significant figures, perhaps you should be writing an article on cost estimating rather than reading one.

For many of us, though, incomplete documentation and cost element uncertainty are commonplace, particularly when new design alternatives are being evaluated and cost estimates play a significant role in the decision process. If you accept these conditions rather than fight them (or worse yet, pretend they don't exist), it seems reasonable to develop a cost estimating technique that accommodates uncertainty in both design and cost elements.

Cultivating uncertainty

One of the problems encountered in estimating from limited information is the reluctance of the estimator to specify a value of which he is unsure. Often his reaction, when forced to give an estimate, is to set the value very high in the mistaken belief that it is always better to overestimate than to underestimate. This problem is reduced if the estimator is allowed to give not only his best estimate, but an indication of the uncertainty associated with the estimate.

Estimates reflecting uncertainty may be specified by three values representing the 10th, 50th, and 90th percentiles of a probability distribution. The 50th percentile is that estimated value which is equally likely to be higher or lower than the actual value. It can be considered the estimator's *best guess*. The 10th percentile value is one for which there is only one chance in ten that the actual value will be lower, and the 90th percentile value is one for which there is only one chance in ten that the actual value will be greater. The 10th and 90th percentile values are roughly equivalent to the estimator's "best case" and "worse case" estimates.

This technique permits the estimator to quantify his uncertainty so that the risks involved in making decisions based on his information can be evaluated. It permits estimating with any amount of information, however limited it may be, since the uncertainty is also stated. Uncertainty is present in all information; even "actual costs" contain some degree of uncertainty.

Narrowing the estimate

When little is known about the product to be estimated, it is tempting to give an overall estimate with a wide range of uncertainty. However, it is almost always possible to reduce the total uncertainty if the product is broken into as many elements as can reasonably be evaluated separately.

To illustrate, assume a product is comprised of two elements for which we estimate costs as shown here:

Element	Cost range		
	0.10	0.50	0.90
A	$8	$10	$12
B	$4	$ 5	$ 7

223

Jim D. Burch is a Manufacturing Resource Planning Specialist at Honeywell Information Systems. His experience with computer manufacturing includes prior positions as Manager of Manufacturing Project Engineering and Manager of Machine Shop Planning.

Mr. Burch is a member of the Society of Manufacturing Engineers, in which he has served as Chapter and Zone Chairman.

Figure 1. Estimated labor and material costs for a product over a five-year period. Note the uncertainty between high and low estimates increases as costs are projected farther into the future.

```
              MANUFACTURING ADMINISTRATION & RESOURCE PLANNING
                        COST/RISK ANALYSIS-$$$
                              08/22/74

                              PRODUCT X


              PERCENT-
                ILE      1974     1975     1976     1977     1978

                 10      2498     2703     2930     3171     3434
ASSY. LABOR      50      2594     2807     3042     3293     3566
                 90      2692     2913     3157     3417     3701

                 10       661      727      794      866      946
TEST LABOR       50       747      821      903      992     1090
                 90       876      963     1066     1180     1307

                 10      7726     7495     7272     7060     6858
MATERIAL         50      8172     7932     7704     7488     7282
                 90      9089     8804     8551     8310     8080

QTY COMPLETE ASSY         117      131      142      143      143


THE FOLLOWING PAGE(S) CONTAIN A SENSITIVITY ANALYSIS OF THE
ABOVE PRODUCT COSTS FOR VARIOUS YEAR(S) IN TERMS OF ITS
SIGNIFICANT ELEMENTS.

COLUMN HEADINGS ARE DEFINED AS FOLLOWS:

  ASSEMBLY = ANALYSIS OF THE ASSEMBLY DOLLARS.
  TEST     = ANALYSIS OF THE TEST DOLLARS.
  MATERIAL = ANALYSIS OF THE MATERIAL DOLLARS.

  NO.   = SIGNIFICANT ELEMENT NUMBER (SEE INDEX FOR DEFINITION).

  LOW   = IDENTIFIES WHAT PERCENT OF THE TOTAL UNCERTAINTY OF
          COST ON THE LOW SIDE , IS CONTRIBUTED BY THE
          SIGNIFICANT ELEMENT.

  MID.  = IDENTIFIES WHAT PERCENT OF THE TOTAL '50 PERCENTILE'
          COST IS CONTRIBUTED BY THE SIGNIFICANT ELEMENT.

  HIGH  = IDENTIFIES WHAT PERCENT OF THE TOTAL UNCERTAINTY OF
          COST ON THE HIGH SIDE , IS CONTRIBUTED BY THE
          SIGNIFICANT ELEMENT.
```

Figure 2. This sensitivity analysis quantifies the cost and uncertainty contribution to each element identified by number in the left-hand column. Note that number 8 has not yet been determined with much accuracy.

1974 SENSITIVITY ANALYSIS FOR PRODUCT X

NO.	ASSEMBLY LOW	MID.	HIGH	TEST LOW	MID.	HIGH	MATERIAL LOW	MID.	HIGH
1	1	2	1	0	0	0	0	2	0
2	1	2	1	1	3	0	0	7	2
3	0	6	0	0	0	0	0	5	21
4	0	12	0	0	0	0	0	3	10
5	0	1	0	0	0	0	0	0	0
8	38	18	36	72	39	72	43	21	1
9	0	0	0	0	0	0	0	1	0
10	0	0	0	0	1	0	0	0	0
14	1	2	4	0	0	0	0	1	0
15	5	3	5	0	0	0	0	0	0
16	0	0	0	0	0	0	0	4	0
18	0	0	0	0	0	0	0	3	0
22	27	15	26	0	0	0	0	1	0
24	3	4	3	0	0	0	31	9	63
27	0	1	0	0	0	0	2	3	0
28	0	1	0	0	0	0	0	2	0
35	0	0	0	0	0	0	0	1	0
37	1	3	1	0	0	0	0	2	0
38	0	0	0	0	0	0	0	3	0
39	1	4	1	0	0	0	0	1	0
41	0	0	0	0	0	0	5	2	1
48	0	2	0	0	0	0	0	0	0
63	0	2	1	0	0	0	0	2	0
64	16	12	15	27	36	27	16	14	1
66	3	4	3	0	0	0	1	6	0
67	1	1	1	0	0	0	0	2	0
168	1	2	1	0	0	0	0	1	0
176	0	1	0	0	0	0	0	1	0

We might conclude, erroneously, that the product cost uncertainty should be expressed as:

0.10	0.50	0.90
$ 8	$10	$12
+4	+5	+7
$12	$15	$19

Actually, if we believe there is only a 0.1 probability of element A costing $8 or less, and a 0.1 probability of element B costing $4 or less, then we infer there is only 0.01 probability that a product comprised of one A and one B will cost $12 or less.

A better method of combining the element costs is similar to one sometimes used in dimensional tolerance analyses. Let's go back to our previous example:

Element	Cost range		
	0.10	0.50	0.90
A	$8	$10	$12
B	$4	$ 5	$ 7

First, we express the 0.10 and 0.90 values as differences from the 0.50 value:

	0.10Δ	0.50	0.90Δ
A	−2	10	+2
B	−1	5	+2

Next, we square the differences and add them. We also add the 0.50 values:

	$(0.10\Delta)^2$	0.50	$(0.90\Delta)^2$
A	4	10	4
B	1	5	4
Totals	5	15	8

Finally, we take the square roots of the summed, squared differences and derive the total product cost range by combining them with the total 0.50 value:

Total product 0.10 value = $15 - \sqrt{5}$ = $12.80
Total product 0.50 value = $15.00
Total product 0.90 value = $15 + \sqrt{8}$ = $17.80

The total product cost uncertainty as determined by this method correctly reflects the fact that it is likely that some of the element cost uncertainties will offset each other. If the estimator has been objective in his determination of the element values, some probably will be higher and some lower than his "best guess." For a product with a large number of elements, the uncertainty of the total cost can be surprisingly small.

Checking the estimate sensitivity

This technique also facilitates further analysis of the sensitivity of the product cost to each element cost. Again, note the example:

Element	Low variance	Best estimate	High variance
A	4	10	4
B	1	5	4
Total	5	15	8

High and low estimates as a percentage of the total estimate are calculated as follows:

Element	Contribution to low uncertainty	Contribution to total cost	Contribution to high uncertainty
A	80%	67%	50%
B	20%	33%	50%

This "sensitivity" analysis may be more useful than the estimate of total cost. For instance, if the estimate is too high, the elements which make the most significant contribution to total cost are easily identified for further investigation as candidates for cost reduction. If the total uncertainty is too great to reach a decision on design alternatives, the elements which contribute most to both high and low uncertainty are also easily identified.

On using this tool...

Although the emphasis of this article is on cost estimating, the technique described here is most effective as part of a larger manufacturing system, which includes labor forecasting and analysis of other critical resource loads.

Samples of the cost estimating output of such a system are shown in Figures 1 and 2. Figure 1 shows the estimated labor and material costs for a product over a five-year period. Costs for each year are estimated as previously described. In this example, element labor costs are projected to increase and material costs to decrease with time. Note also that the uncertainty increases as costs are projected farther into the future.

The sensitivity analysis, Figure 2, quantifies the cost and uncertainty contribution of each element identified by number in the left-hand column. Apparently the cost to test element number 8 has not yet been determined with much accuracy, but it should be recognized that the 72 percent is a relative value and would not be very significant if the total uncertainty were small. Also, note that the estimator is very sure that the material cost of element 8 will not be much higher than his "best guess," but considers it possible to be much lower.

A system utilizing these features has been in operation for over three years with varying degrees of success. Its major advantages are the ability to work with limited product descriptions and any amount of element cost uncertainty. It provides a common base for evaluation of alternatives, permits sensitivity analysis and, when implemented via computer time-sharing, provides very fast response.

The primary disadvantage of this technique is that many people are not yet ready to accept uncertainty. It is obviously much easier to base a decision on comparing two discrete values rather than two overlapping probability distributions. Similarly, many people have difficulty providing probabilistic inputs to the system. An estimator's confidence in his estimate is heavily biased by his perception of the consequences of being wrong. Most people readily accept a 0.5 probability of buying two cups of coffee, but would be very reluctant to flip a coin for $100. It is not unreasonable, therefore, for a shop manager to estimate his "most probable" labor cost to be one he is really 0.99 sure of if he perceives the consequences of overrunning the estimate to have a much more negative impact on his life style than underrunning. Furthermore, he can probably *prevent* an underrun if the consequences are the same.

In spite of these problems, the value of the approach to cost estimating described here outweighs its disadvantages and is gaining acceptance as a useful tool in business planning.

MANAGING A COST ESTIMATING TEAM

7

Towards The Cost Estimating Speciality

by

Phillip F. Ostwald

Cost estimating is the unrecognized field of management. This surprising fact emerges from a survey of over 100 companies. With over 400,000 firms in the United States producing a product, the cost estimating function is mandatory. Despite the few academic courses and textbooks promoting this professional field, cost estimating is prominent in management.

A diversified list of 104 companies were questioned to determine the patterns of organization, staffing, policy formulation, and work content of the estimating function. The list included job shops, moderate to continuous production, and small, medium and large companies. The distribution of the survey is shown by Figure 1. This profile, as identified by the Standard Industrial Classification, would lead to differing conclusions, if for example, the chemical industries dominated the survey instead of fabricated metal products. Governmental agencies have been censored from the list although there are some federal departments that estimate services and products for "selling" to other governmental departments.

BACKGROUND OF COST ESTIMATORS

The education for cost estimators varies. He or she is normally trained in the engineering or accounting field. For technology-oriented products, engineering is more common. Some estimators have received shop training only, and in view of the requirements for practical knowledge, a department staffed with academic and shop backgrounds is believed superior to one with only one kind. These results are shown in Figure 2.

It is not surprising that small companies train their estimators almost totally within the organization. Professional association and development of the estimator's identity has little opportunity to flourish. Up-through-the-ranks is more common than formal education.

Once an organization is able to hire time study technicians or engineers, cost estimating becomes more formalized. Even though time study and predetermined motion time data experiences are helpful for estimating direct labor, an estimating philosophy developed on that background alone is often counterproductive unless complemented by other professionals within the organization.

If the organization is larger than the time study threshold, professionals with backgrounds in industrial engineering, manufacturing, accounting, design engineering, and purchasing introduce their disciplines, and the organization blends these ideologies. Oftentimes the professionals are re-titled as cost estimator or cost engineer. Even within larger organizations we find shop-trained people practicing cost estimating. As estimating requires an intimate experience with the manufacturing processes and the products, estimators tend to be older than their counterpart professional. For fear of making critical mistakes, few junior professionals are found in the cost estimating function.

---

copyright, Phillip F. Ostwald

RESPONSIBILITY AND ORGANIZATION

Viewed internally within the organization, responsibility is here defined as the obligation of a cost estimator, to whom a superior has assigned a duty, to perform the services required. Where within the organization is this duty performed? Figure 3 gives the results of the survey and shows that for the SIC profile, manufacturing dominates as the most likely source for the cost estimating department.

There were hidden nuances to the organization question that tell as much, however. While a manufacturing division had control, it was the industrial and manufacturing engineering department that provided the implementation role. Under this leadership, a numerical approach to the finding of cost values was commonplace. Emphasis upon labor standards, manufacturing routing, tooling, and material calculation were noticeable factors in the calculation of direct costs.

Similarly, when the cost estimating function was under the direction of accounting, emphasis was more likely to be upon a standard cost approach, as typified by these numerous systems, and then achieving compliance a la management pressures and control procedures.

If marketing or sales managed the cost estimating activity, emphasis in cost estimating tended to rely on historical analysis of past costs. Sometimes there would be leverage to artificially reduce real costs so as to promote sales. Difficulties in keeping pricing and cost estimating distinct were more apparent whenever sales had authority for cost estimating.

While the reporting lines for cost estimating may be seen on an organization chart, the cost estimate document usually ended up in the in-basket of a senior executive for final approval, indicating its significance to the firm. In the small company, it was the owner or president who signed off. Even in medium to large organizations, the president, vice president, or a senior executive accepted the terminal authority. For the largest corporations a plant correspondent would provide local information to support the total corporate requirement, but final approval by a senior executive at the plant was customary. It is interesting to note that sign-off approval for final implementation resided with higher executives for cost estimates than for other routine manufacturing documents.

STAFFING AND DEPARTMENTAL STATUS

The number of cost estimators varied with many factors, but Figures 4 and 5 summarize the findings. Figure 4 gives the average number of estimators the number of employees at a plant site. Viewed differently, Figure 5 indicates annual sales versus staffing. It is evident that the number of estimators are few. While all product producing companies have the estimating function, there were some that required the duties of an estimator only part time. For the small company, it may be the president or another senior executive. As the companies become medium-sized or larger, the function is recognized and job analysts can list the activities that describe the estimator. Job titles reflect this specialization.

If a department titled "Cost Estimating" exists, it is usually under the sponsorship of industrial engineering. Accounting, though it recognizes the requirement for a cost estimating department, accepted the department without preemptive right to the word cost. Moreover, accountants are supporters of the identified organization even though it is not under accounting leadership. If estimating is located under cost accounting, it tends to become a coordinative activity except for the application of overhead and other indirect costs. Not all companies organize the cost estimating activity into a department, preferring instead to recognize it as a function and a part of another department.

If there is a Cost Estimating Department, its behavior is unlike the routing, or methods, or accounting department for there are few models that have been identified. Initially, cost estimating allows the department to be centralized or decentralized. In the context of a centralized operation, a department tends to be self reliant, that is, it develops standards and standard time data, estimates labor and materials using the bill of material, and ultimately determines bottom-line cost for the product. For example, a centralized group would estimate tooling cost itself rather than request that information from another source. Contrasted to this style is a department that depends upon other groups for information. Using requested information, the cost estimating department processes the information into a bottom-line cost. The result of both management styles is the cost estimate document or "recap." It should be pointed out that few departments operate at either extreme, but there is the tendency to act centrally or not. Figure 6 shows the results of the survey. With the exception of smaller firms, where a centralized operation would be expected, the results are mixed.

Decentralized cost estimating is normal for large and separated corporations, or ones where long-time bidding periods are found, or the high-technology groups. Weapon system contractors fall into this style. Mature organizations, machine tool companies, OEM, and geographically compact firms tend to have centralized organizations. In some organizations it is mixed, as a centralized group estimates new products, yet the re-estimates, repair, and salvage estimating may be parcelled to the centers incurring the cost. Thus, generalizations on this matter do understate the variety of organization.

Small companies do not have the difficulties attendant with the question of how to organize cost estimating. With the medium-to-large companies, there are nagging problems that become apparent in the inability to cost estimate accurately and consistently for products or within product lines. These organizations have not developed instruction manuals or a pro-forma method in assembling the variety of costs into understandable or convincing documents. The large corporate groups seemingly have overcome these difficulties.

These inherent management difficulties have led to the new concept of the Cost Estimating Committee as advocated by the author. The committee approach is a form of decentralized cost estimating, but its impact is difficult to judge at this time.

COST ELEMENTS

What cost elements are estimated? Our survey summarizes these findings as Figure 7. For the purposes of the survey we classified key costs and then assessed the contribution by estimating.

In elementary situations, the estimate was calculated using the direct labor hour, that is, all direct and indirect costs (with exception of direct materials) were lumped into a direct hour rate. Estimating the number of direct hours and then multiplying by a direct hour rate and adding the material costs results in the full cost of the product. This, we concluded, implied that estimating determined the future value and all of the cost elements as shown on Figure 7, if an approach of this kind was used.

Contrasted to this approach is one where companion estimates and procedures were identified for each cost element, direct labor, direct materials, tooling, overhead, selling, and engineering, etc., and the cost element is compiled and estimated separately. Ultimately they are gathered using arithmetic rules and formal processes to arrive at a bottom-line cost. If the estimating department determined the future value of the cost elements as shown on Figure 7, the survey entry was tallied "Yes."

It is pointed out that these elements, in one way or another, are always determined by the "function", which is inclusive of the department. But these elements may not always be contributed by the department. For instance, about 30% of the time the cost estimating department would add profit to the cost thus converting the value to price. In the other 70% of cases, the profit-to-price conversion would be done elsewhere in the organization.

The survey shows that direct labor and direct material are almost always a part of the estimator's work. In those cases where it is not, the sales department dominated the efforts, and the approach is more attuned to price setting, not estimating. The matter of overhead is mixed. The allocation of indirect expenses, unitizing, and finding the rate is a cost accounting responsibility, but the application to the estimate is often a part of estimator's work, due to convenience and other reasons. Where estimating did not apply the overhead rate, the estimating documents are transferred to accounting for their continuation. Usually the contribution of cost elements by the estimators does not involve the addition of profit or the determination of price except for the smaller firms. The price-setters or profit-groups of a firm are specialists, and whenever the cost estimating and profit/price setting becomes intermixed, one activity appeared to suffer. Thus, price setting is often delegated to non-estimators.

REFERENCE

Ostwald, Phillip F., <u>Cost Estimating for Engineering and Management</u>, Prentice-Hall, Inc., 1974.

STANDARD INDUSTRIAL CLASSIFICATION	PERCENTAGE
Fabricated Metal Products	35 %
Machinery, except electrical	16 %
Electrical machinery, equipment and supplies	15 %
Ordnance and accessories	8 %
Chemical and allied products	5 %
Other (Food, lumber, paper, rubber, glass, primary metals and instruments companies)	21 %
	100 %

Figure 1 DISTRIBUTION OF SURVEY

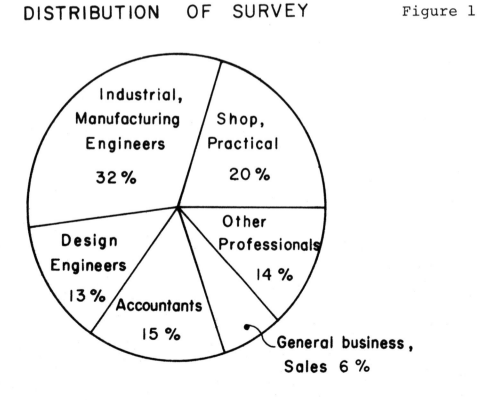

Figure 2 Background of Estimators

Figure 3

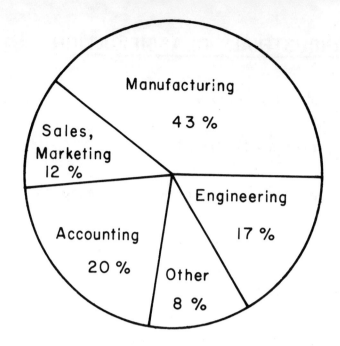

Responsibility for Cost Estimating

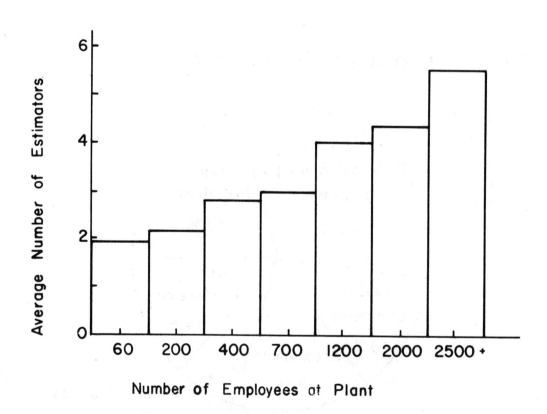

Figure 4

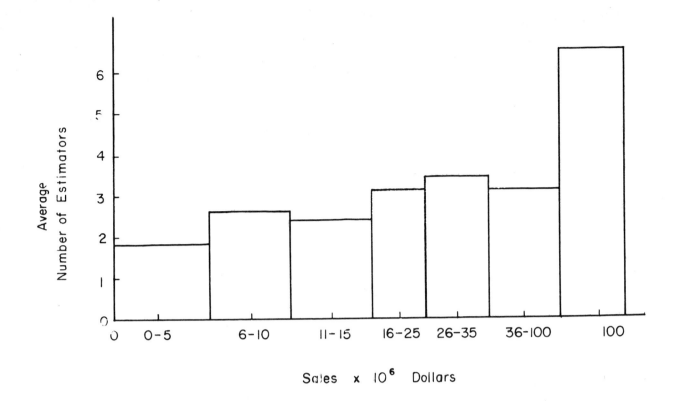

Figure 5

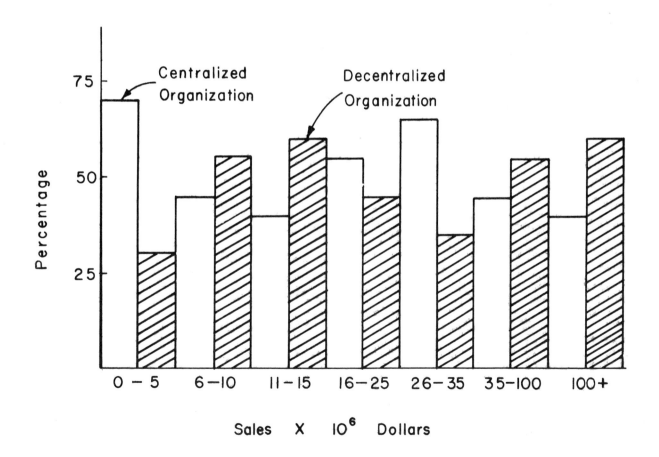

Figure 6

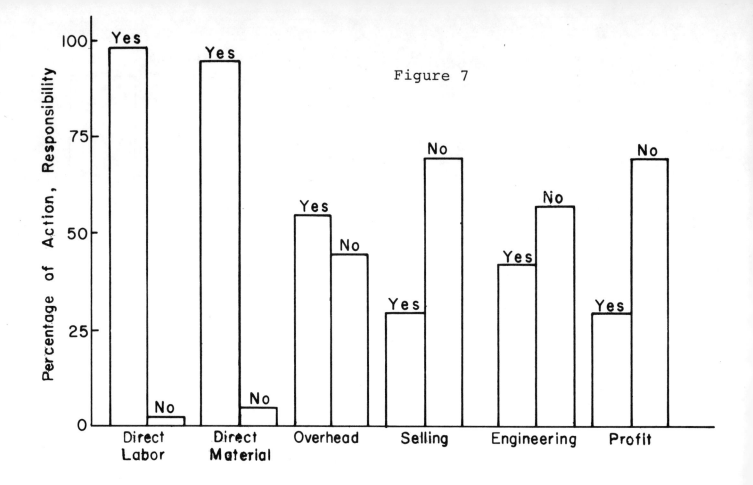

Figure 7

Cost Elements in a Product Estimate

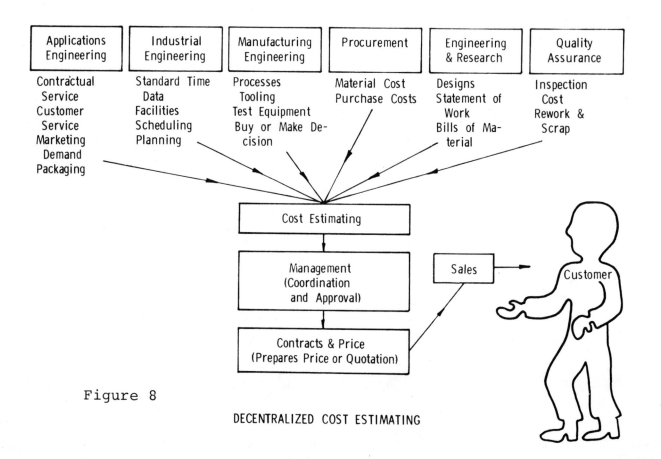

Figure 8

DECENTRALIZED COST ESTIMATING

Top Management And The Estimating Function

By Robert C. Berman
President
Berman and Associates, Inc.

A quality job by the estimating function depends on both Cost Estimating and Top Management fulfilling their obligations to each other. Some Estimating obligations are frequently not stressed: judgment about detail; organization for speed; reflection of trends and programs; product definition assessment; realism about probabilities; manipulating weak cost systems. Conversely, Top Management's obligations are almost universally ignored, despite the fact they have a direct bearing on the estimator's effectiveness. Management is reminded of them here: explain the purpose; share plans and interpretations; spread respect for Estimating and support it; remove reported roadblocks; and trust the estimator's thoroughness.

INTRODUCTION

I used to consult with a company that went through the annual ritual called "Five Year Planning". I remember the President once received a Division's plan which forecast a very meager profit for the following year, based on several items of unavoidable bad news: aggressive new competitors in the market, pre-negotiated labor rate increases due at specific dates, new product development delays, and forthcoming unfavorable legislation.

That President "rejected" the plan as "unacceptable". He forced the Division Manager to go back and re-do his plan so it would forecast a better profit. And he was right to do so --- there was a lot of fat and no counter strategy in it. When the revision arrived, now honestly lean and positive, he still declared its bottom line to be "unacceptable". So it was done again. And again. Those last two times the President wasn't right.

Finally a P&L plan arrived showing a delicious return, and he "bought" it. By that time it was totally impossible to achieve, of course. But it gave him the answer he had decided, in advance, would be acceptable.

A year later the President was aghast when the Division's results came in bad --- as bad as the original revision had forecast. Not only did he hate bad surprises, but he felt actually betrayed by the executive who had "lied" to him. In fact, he fired him.

The President never did recognize whose fault the "surprise" was.

Why have I related this story? Because this is going to be a paper dealing with some of the interfaces between top management and the cost estimator. And I'll venture to guess that an exactly parallel situation occurs daily somewhere in the country, where some misguided top executive sends his estimator back to "do it over" when he's presented with a cost estimate whose total he doesn't like to see.

What should the estimator do if he is faced with such a situation? Assuming his original submission was the way he honestly saw it, should he dig in his heels and refuse to prostitute his profession? And thereby risk his job? (After all, not only could he be judged immature for not appreciating the pressures imposed by low-cost competitors, but -- like the king's messenger in olden times -- he could be beheaded merely because he's the bearer of bad tidings!)

Or should he save his neck at the time, pull all the realism out of his calculations, return with an impossibly-optimistic revision, and risk his neck (and, incidentally, the president's) later, when every product sale priced on the basis of his estimate gives away a piece of the company?

The actual fact is: in real life he'll often do the latter. After all, future losses might be blamed on somebody else. Also, maybe we really won't need that contingency factor if (miraculously for the first time in the company's history) everything runs without a hitch.

What's more: a job in the hand is worth two in the newspaper.

I was mulling over that sad situation when I started this paper. It occurred to me that an awful lot has been written about the responsibilities of you estimators and what you owe to your superiors. But nothing is ever said about what your superiors owe to you, if they expect to get the best performance from you.

So I decided to deal with both. Half of my time I'll be talking to you cost estimators who are here in the audience, and I hope I can give you some new views about what you are (or aren't) doing. The other half of my time I'll be addressing somebody who is probably not here --- your president. I am counting on you to take home a copy of this paper and get it onto his desk. If he's as big as most of the executives I work with, he will appreciate being reminded of basics he may be forgetting.

ASPECTS OF ESTIMATING

First, then, some words about your own work. I want to cover six of your obligations which are seldom stressed, and sometimes are even unrecognized.

An orderly way to do this is to tie them to my own definition of a cost estimate. So if you'll please write this down, you can refer to it as we go: "A cost estimate should be your CLOSEST TIMELY PROJECTION OF SOMETHING'S MOST PROBABLE FUTURE COST." Almost every word in that definition touches on some responsibility you have to top management. Let's examine them:

"...CLOSEST..."

This calls for your ability to strike a sensible balance between extreme accuracy and roundhouse approximations. Your skill here is one of judgment: when is a number close enough to be used without playing around with it any further?

An unskilled clerk could do your job if all your figures came from factual hard data. But of course they don't. So some one like you is relied upon to combine a few facts with a few educated guesses, to know when an educated guess must be mighty close or when it could be off and not really matter.

Further, you must have the maturity to lay aside, sometimes, your engineering mania for precision when it doesn't make sense. For example, calculating a unit direct labor cost to the nearest 1/100th of a cent is silly if you then go on to expand it by an approximated 22% fringe allowance or a global 250% plant-wide overhead factor. That would be equivalent to estimating the distance to some star by measuring from your nose to your office door in millimeters, then guessing the rest in approximate light years.

"...TIMELY..."

This calls for your ability to produce estimates "yesterday". I can't think of many occasions when a cost estimate isn't asked for "ASAP". Perhaps you are allowed some reasonable time if the estimate is to be used in a capital investment justification, or to influence a make/buy reversal. But if it is needed to help the top man set or quote a price, you can bet that time is important. Your company may be racing against a deadline for a bid, or the date set for introducing a new product at the annual trade show. It's up to you to respond quickly, including plenty of homework if necessary, and to assume that when the president says "rush" he has a good reason.

So part of your skill is knowing exactly what information you need and also knowing where to get it. Therefore it follows that part of your obligation is constructing files and forms or checkoff sheets or plug-in formulas or EDP programs that will let you massage it all into a total estimate as quickly as possible. If you haven't yet organized yourself in this way, you still have an open job to do when you get home.

"...PROJECTION..."

This calls for your imagination. You are giving somebody a picture of the future, and you are responsible for considering <u>everything</u> in that future which might affect the cost. I won't dwell on the obvious things, more than to list a few here: labor rates and purchased material costs will undoubtedly be higher, as will the costs of tooling and equipment; certain overhead items will cost more, but some may cost less; efficiency may be higher; scrap rates may be lower. Consideration of these is automatic.

But the skill which will really set you apart is being able to quantify the progress and future levels of company and national programs and policy changes. If the product you are estimating will be made during the period just preceding a labor contract expiration, what will productivity be then? If the Quality Control function is being strengthened, what will be the impact on yields, rework, scrap? If old equipment will be older, will there be more downtime? Or if a preventive maintenance program is being installed, will there be less downtime? If gasoline prices soar, will you be paying more for freight-in on purchased items? If environmental codes are closing foundries by the dozen, will casting costs go through the roof? Will financially-troubled cities raise payroll and inventory taxes? What will be tomorrow's government impositions with regard to vacations, holidays, pensions, workweeks, insurance, employee safety, product performance, depreciation methods? How about the company's reorganization plan, which will raise or lower the number of locations or indirect staff?

Enough! You cannot be expected to know everything about everything. But I do believe you have a duty to stay abreast of the times, inside the company and out, and crank into your estimate as many of these trends as you can, once you foresee they'll have a cost impact.

"...SOMETHING..."

This calls for your familiarity with the item being estimated, and your ability to recognize its similarity to (or difference from) a comparable item for which some data are available. If you haven't set up some kind of grid reference system to help you extract pieces of old estimates from your files, then you are making it hard for yourself to fulfill this duty.

Also implied is your ability to recognize if the "something" hasn't been sufficiently defined for you in the first place. After all, you can't estimate an assembly without even a rough list of its components. You can't estimate the cost of sheet steel without knowing its required gauge. It is your job to know if you haven't been told enough. And it is your responsibility to insist on getting the missing definitions before you put a cost tag on the whole.

There is a bit more to this "something" as well. There is the matter of your opportunity to contribute to design. Very often you'll be asked to estimate a product which you immediately recognize is a costlier design than some similar one produced long ago. Now, I don't say you are responsible for challenging the designers --- you have enough to do just evaluating whatever they give you to estimate. You'd have every right to take it as it is, shrug at their stupidity, and then proceed with the costing. But I say shame on you if you ever get an improvement idea and don't bring it up! Especially when you're in possession of such a beautiful "memory file".

"...MOST PROBABLE..."

This calls for your sense of realism. The biggest trap an estimator can fall into is believing the optimists and egotists. Your circle of contacts (including the president himself who is hoping for a low-cost answer) is full of purchasing men who are sure they can negotiate bargains, production men who expect to break output records, schedulers who anticipate no conflicts, designers who release only perfect prints, and executives who are going to under-run all their budgets.

You just have to assess who is right! If we have been buying X percent of our goods at premium rates because we never allow enough vendor lead time, why won't this continue? If the workforce has averaged 82% of standard for the last five years, what will get them up to 100% by August? If the last week of each month is always a panic of overtime in order to hit shipping targets, why should your estimate assume only straight-time costs? If one third of all rework is traceable to corrective design changes, what magic will suddenly make the engineers turn out flawless documentation?

You cannot and should not assume the worst of all worlds, or you'll cause every product to be overpriced and no business will come in at all. But you must reflect what will most probably happen. Or else the company will take in a lot of loss sales, which is just as quick a way to die.

"...FUTURE COST..."

This calls for your ability to handle the company's existing cost collection system. I've already dealt with the "future" aspect. But a lot of your estimating starts with records of present cost and then modifies from that base. So you must be alert to system deficiencies which might make a present cost unreliable to use as is.

For example, if setup time on a machine is not recorded separately, but is combined with run time so the total of the two is charged to the pieces made, watch out! ---the recorded cost per piece for a run of 6 will be far different from a run of 60, and an average of both will of course be meaningless. Or, to take another case: if the cost to run a piece through five successive operations is accumulated in a shop order, then divided by the finished quantity to get a cost per piece, watch out! --- any pieces scrapped after operations #1 and #2 will inflate the time numerator, but won't show up in the quantity denominator at all. If downtime and rework are charged to the job, watch out! If the only available record is the last run made, watch out! And if overhead is applied as one massive percent of labor, with no breakdown -- either by fixed vs. variable, or by plant department, or by product line, or by machining center, or by something -- watch out! watch out! watch out!

So I am saying: if the existing cost accounting system doesn't give you reliable present costs to work with, it is your job to know it, to scream about it, and (until it is fixed) to develop your own alternative data for use in estimating. If this requires you to take some more Accounting courses, what's wrong with that?

TOP MANAGEMENT'S ROLE

"Oh boy," I can hear you thinking. "If I really tried to do all those things, I'd have the world's biggest frustration ulcer by tomorrow morning. In fact, just hearing about them, I think I feel it breaking through right now! Because no way do I have the authority, the information, or the credibility I'd need to give top management what you say I owe to it."

Probably true. But there's hope. Because now I am going to talk to your top management. I shall try to point out what they must do to clear your path, so you can get on with the job as I described it.

PURPOSE

The earliest thing management owes to the estimator is the reason why a particular estimate is being requested. It does make a difference. It will help the estimator judge the degree of accuracy required. If he can do that well, he will neither kill himself with detail when "close enough" is good enough, nor will he overgeneralize when a major commitment hangs on the balance of a penny or two. I am currently working with a chemical company, for example, where a few cents per pound of one ingredient in a formula carries an annual profit impact of $250,000. In that same company, a decision to close a plant and totally subcontract its work was influenced by item estimates to the nearest $5000, because all management really needed was an order of magnitude to point the direction.

If the estimator knows the purpose to which his numbers will be put, he'll also be able to prioritize with some sense. You can bet he has a backlog, and you can bet every request in it is labeled "RUSH". But some rushes are rushier than others (how is that for communication?). Even if the estimator is willing to work his heart out on overtime, he will be helped (and spurred, too) by the knowledge of what each estimate will be used for next.

PLANS

Management owes the estimator as much as it can securely tell him about upcoming company programs, restructurings, and changes to "what has always been". After all, he is forced to be guided by records of past and present costs. If management plans to change something, the estimator has to know when that will make some historic cost totally invalid, or valid only if factored up or down. The cost estimator has a responsibility to nose around for such news, and I have already covered that. But top management owes him help with that detective work, in the form of information and warnings about cost-related company moves yet to come.

Similarly, executives are more likely to know about the outside business world's trends --- industry, the laws, labor relations, international supplies, etc. If they learn of something at the Chamber of Commerce luncheon or the trade association annual meeting, they should pass it along to the estimator who wasn't there. He can digest his newspapers hungrily, and he should, but special trend interpretations still are best made by the top man himself.

STATURE

Not the first listed, but the greatest by far, is the need for top management to ensure an organizational respect for the estimating job. It must be made clear to all departments that estimators guide executive decisions, directly affect pricing and profit, and therefore need willing and quick cooperation when they come looking for information.

After all, these men have to contact almost every other department as they put their numbers together --- Purchasing for quotes, Accounting for costs, Engineering for specs, Personnel for rates, Scheduling for dates, to name a few. It would be amazing if they weren't resented when they appear, because they mean an interruption in the "much more important" regular work to be done. So they encounter impatient, peremptory treatment. They irritate people if they persist. They enrage people if they challenge offhand answers. They are viewed as job security threats if they seek data which only one man jealously knows. Or if the true answers to their questions will expose somebody's poor performance. Or if they weed out the contingency factor that each informant tries to build in.

The calibre of the estimator's final output, as well as the speed with which he can deliver it, depend greatly on how other people view the importance of his work. Spreading appreciation of that importance is directly the responsibility of management. The top man must make sure his estimator has access to the needed information.

SUPPORT

Merely telling everybody to cooperate will not guarantee cooperation. You will have resisters; there will be challengers to test whether or not top management really means it. The day will come when Estimating refuses to estimate until Engineering produces better product definition, and Engineering will claim it has neither time nor budget to expand upon information that "should be enough" if the estimator only knew how to do his job. When these conflicts arise, presidential support is needed --- quickly and unequivocally. Without it, prior instructions about cooperation will be exposed as meaningless prattle, and the cost estimator will thereafter be powerless to deliver the work as he should.

RESPONSE

Top management owes <u>all</u> employees a critical but attentive ear to ideas for improvement of the enterprise. It also owes them action on the good ones. Obviously, management owes this to itself as well -- a kind of selfishness.

In the daily work of the estimator, he is likely to identify many areas where changes in other departments or systems would improve his own effectiveness. He doesn't have the authority to make such changes, of course. But the top man does. When the estimator brings these up and proves they are valid, the executive owes him response by directing that weaknesses be corrected and improvements be instituted.

I've mentioned some of these examples earlier: inaccuracies in the cost collection system; overhead multipliers so broad as to be distortive; design change notices not fully distributed; failure to include Estimating in pre-release reviews, customer-requested changes, or notification of shop methods improvements and equipment additions.

Where cures are interdepartmental, only top management can effect them.

CONFIDENCE

Assuming the cost estimator has already proven his skill and reliability, top management owes him acceptance of his numbers. Not without a review; the estimator should be allowed to, even expected to, comment on his results and the premises underlying them. Not without a review, but without a hassle. The president shouldn't say, "But Joe insists that this time he could beat the standard." Or, "Surely the new machine will be de-bugged by then." He should believe that these things have been considered.

Or, worst of all, "Ye Gods -- with costs like <u>yours</u> we'd never be able to sell anything around here!" The implication is that high costs are the estimator's fault, and he should go back to his calculator and personally conduct a cost-reduction program. Wrong! He is not a cost creator, but a cost forecaster. He is not a pricer, but a guarantor that management's eyes are fully open while management sets the price. If the estimator's numbers show that costs are too high, top management's response must be to fix the causes, not harass the reporter.

You'll recognize that this is where we started. So, having circled around to it, let's call it the best place to end.

I hope I have given you at least a few new thoughts about the interplay between top management and the estimating function.

Fabrication and Assembly Cost Estimating

By Arnold M. Kriegler
Manager, Fabrication Engineering
Collins Radio Company

It was a bit presumptuous of me to accept the invitation to talk to you about estimating, because until 2 years ago, I had never had any direct responsibility for product estimating. But shortly after inheriting an estimating department as part of a fabrication engineering responsibility, this opportunity arose; and since I was in the process of trying to analyze these new responsibilities, this seemed a good way to approach that task while hopefully gathering information that might prove of value to you. So, what follows is a neophyte's judgment of the estimating task and what it takes to do it competently.

Well, it didn't take much analysis to discover that it's a business that requires a certain amount of skill and daring. Picture this: The marketing man cultivates the customer for months, assuring him that his desire for unique or peculiar product features presents no problem to the inventive genius and the manufacturing know-how of the Company. He then parleys with the designers, assuring them that we must provide the new features to clinch the order; that there is no way, absolutely no way to sell an off-the-shelf item. More time passes, until the designer announces that the new features can indeed be achieved, and that he has the "design concept" well in mind.

Now, while all this has been going on, the estimating world has been blissfully ignorant of the opportunity that is about to be thrust upon it; for the customers' procurement lead time has been ticking away during the customer-marketing-designer negotiation, and there will now surely be a crash requirement for an accurate cost and tooling estimate based on the aforementioned "design concept," and sketchy information about the ultimate production quantity, or the production rate. In fact, the only thing that seems known for sure is that the first deliveries will have to be made in about one-half normal lead time, and the bid in no time at all.

I suppose that picture seems exaggerated to some of you, and it probably has a lot of manufacturing bias in it, but I'm sure many of you recognized it. Let me translate that picture into more concrete terms to better define the total estimating task to which my group contributes the fabrication portion. We are called on to render estimates for over a thousand bids per year for state-of-the art electronic equipment. The "design concept" I referred to might be blackboard sketches, yellow pad sketches, preliminary design drawings, or well-defined working drawings, depending on the maturity of the development project. We might have access to a working engineering evaluation unit, or a balsa wood model, or no model at all. It can involve a few parts, for example, a modification kit, or anything up to thousands of parts for a communication system.

The ultimate quantity of equipments to be produced can range from one unit to five thousand or more units. This factor of course requires that the estimator be aware of many alternate processes and tooling methods. To add to the challenge, an average of only four working days is available for our fabrication piece part estimates, according to a sampling of the last year's activity. The average total time available for a bid; that is, the dissemination of bid requirements, generation of a current list of material, estimation of new parts and assemblies, and various bid reviews is about 3 weeks. The range is 2 days to 60 days.

Well, through the eyes of this relative newcomer to the estimating business, those statistics represent a formidable task, and I would like to submit to you some principles that seem basic, and which contribute significantly to accomplishing that task.

The principles I will discuss are these:

1. Define and document all procedures, formulas, estimating data, and responsibilities.
2. Appoint only experienced, qualified people as estimators--avoid the pitfall of shunting marginal producers into this group.
3. Create a system for rapid, accurate retrieval of past cost data.
4. Treat detailed cost estimating as separate from pricing activities.
5. Measure estimate quality, schedule achievement, and cost.
6. Provide for reviews of estimates with various mangement levels.

These principles, I believe can be applied effectively to any estimating activity. Here's how they are applied in the environment I described a few minutes ago; over a thousand estimates per year, quantities of one unit to over five thousand units, complex state-of-the-art equipment, various levels of design definitions, all resulting in an annual sales volume of over 200 million dollars.

Principle 1. <u>Document</u> <u>all</u> <u>procedures</u>, <u>formulas</u>, <u>estimating data</u>, <u>and responsibilities</u>. A glance at figure 1 will give you some insight as to why clear-cut rules

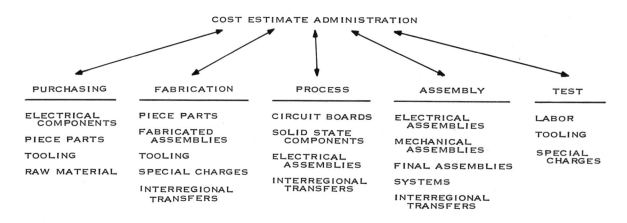

Figure 1. Estimating Responsibilities.

and regulations are important to us. With about sixty estimates in process at any one time, and the fact that we buy and sell components and assemblies between our geographically separate regions, it is clear that some precautions against chaos must be taken. This principle is fulfilled by three techniques.

First, a Formal Cost Estimating Manual, which is an integral part of our Company's Industrial Engineering Standard Practice Manual, specifies the proper use of labor factors such as learning allowances, personal allowances, manufacturing allowances, standard cost formulas, and labor rates. It defines product factors such as complexity, lead time, and design stability. Its use is essential to consistent estimates which in turn reduce the burden of audit support required by many of our contracts. Most of you are aware of the numerous features of government contracts which make audit support documentation necessary and desirable; renegotiation, cost plus fixed fee, cost plus incentive fee, and value engineering clauses, to mention a few. The job of the auditor, and our need to spend time with him, is greatly reduced by the fact that we have a focal document which he understands and can check our estimate development against.

The second technique is one of providing a coordinating function. Each bid request becomes the responsibility of one of ten cost estimate administrators. These men are attached to the Manufacturing Cost Analysis Department, which is separate from any department having a responsibility for detailed cost estimating. These administrators, who generally have degrees or background in Business Administration, Accounting, or Engineering, disseminate bid information to the individual estimating areas, schedule bill of material release dates, estimate deadlines, assemble the bid package, and conduct the various bid reviews. Some 80 to 100 bids per month are coordinated by this group, and a daily status report is rendered for a management overview (figure 2). Their efforts assure that we don't get caught in our own web, and provide valuable experience for them.

BID REQ NUMBER	EQUIPMENT	BID TYPE	PR DP	QTY	VOL	REV DATE	OPEN DATE	CN AD	ES AD
8642-01	331D2-D/333F-1	NFP	16	83	50K	06-12	06-16	DD	KR
6672	980H-6/878L-19A/28/29/30	COMM	11	15	95K	06-12	06-12	RP	ER
8801/02	APOLLO REPAIR	CPFF	15	2	10K	06-12	06-13	LB	FF
HUDEK	TSC-60 COUPLER	PLAN	11	130	1M	06-13	06-13	VH	RF
6692	51Y-7	FP	13	8125	22M	06-13	06-27	RP	RB
8862	SRA-33 SPARES	FP	38	218	1K	06-17	06-17	KD	GS
HUDEK	ARN-58	PLAN	12			06-18	06-18	VH	RF
HUDEK	AS-909	PLAN	12			06-18	06-18	VH	RF
8542	2KW-VHF TRANS	FFP	11	5	170K	06-18	06-23	NG	RH
6697	AN/GRC-153A	FFP	11	20	300K	06-20	06-25	HK	EB
6684	618M-2B MOD KIT	COMM	13	500	10K	06-23	06-23	ES	KR
6687	C-8500/ATE	COMM	15	VAR	50M	06-23	06-23	JH	ER
8874	ARC-51BX SPARES	FP	38	2	1K	06-24	06-24	KD	GS
8894	661F-1	FFP	12	10	75K	06-24	06-25	WS	RB

Figure 2. Daily Status Report.

The third technique is a discipline which states that for a given part number, one and only one unit of the Company is responsible for its production or procurement. This is accomplished by means of an Action Responsiblility Code which is assigned each part or assembly as the design is released. The code designates a particular buyer if the part/assembly is to be purchased. Of course, we have the flexibility to change the Action Responsibility Code, but this is done according to rigid procedures. The use of this discipline greatly augments the power of computer programs which deal with our tens of thousands of active part numbers. For example, each production control area and each buyer receives a net requirement list each Monday displaying those parts or assemblies for which he is responsible, complete with schedule information. I think it is obvious that this is a very powerful tool, and the same capability that directs a part number to the correct production or purchasing responsibility is also used to direct it unerringly to the corresponding estimating responsibility (figure 3).

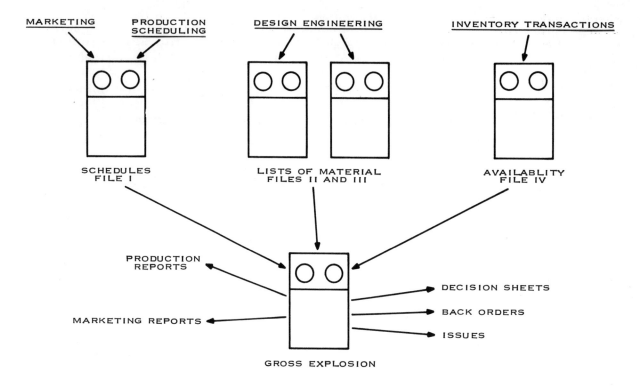

Figure 3. Generalized Material Control System.

Those three techniques, the cost estimating manual, the cost estimating administrator, and the action responsibility code are the prime tools which serve to define and document cost estimating procedures and responsibilities.

Principle 2. <u>Appoint only experienced, qualified people as estimators</u>. I suppose this is so obvious that it doesn't require much explanation, but it is included here in the interest of completeness, and because I believe how you keep good estimators from getting bored or stagnant is worth discussing.

We recently combined the fabrication division's cost estimating and value engineering functions under one department head, and have gained in several ways. We are better able to meet those crash deadlines by being able to shift capable manpower freely between these functions which require very similar skills. We have an increased sensitivity to cost reduction opportunities on the part of the estimators; and just as important, I believe we have added a little variety to their work diet. And, after all, who has a better awareness of new products coming along, and a better opportunity to suggest changes before the design is frozen.

Principle 3. <u>Create a system for rapid, accurate retrieval of cost information</u>. This ability probably means the difference between success and failure in producing numerous good estimates on short lead time at reasonable cost. We have a number of things going for us which make our task easier.

First of all, we use a standard cost system which gathers labor, material, and overhead costs on each existing part number or assembly. The basic information for this file comes from purchase orders in the case of purchased parts, or from the fabrication, or assembly time standard generated by the department responsible for producing the parts or assembly. This means that we can go to this file, either by use of summary tabulations, or through data processing for accurate, up-to-date cost information on any of the almost quarter-million part numbers already in existence. This file is also useful for reference in the case of parts/assemblies which have not been released, but which the designer recognizes as "similar to" an existing part/assembly.

Another key file is the master bill of material file which contains complete part number content of all deliverable systems, equipments, or modules. We are going to talk about the use of these computer files for estimating purposes, but I am sure you recognize that they are used for many other tasks including the powerful programs mentioned previously. Let's see a couple of ways to put these two base files to work.

Suppose that we just want to bid an existing equipment. That sounds simple enough; just peek and see what was bid before. But many of you are faced with the same situations we have--if many weeks have elapsed since the last bid, there have been engineering revisions, methods improvements, quantity changes on component parts, time standard changes, labor rate changes, and perhaps even allowance changes. In order to be sure that these changes are reflected in the estimate, we ought to start all over, and we can do just that through the two base files we are talking about.

As the changes occur, it is a matter of course that the files are updated to reflect the latest information. What we do then in the rebid situation is to marry the current bill of material file to the standard cost file via an estimating application program, and get as output a priced-out bill of material in which we can have a high degree of confidence (figure 4). With a modest amount of manipulation for quantity

REPT 32996 DATE 0-23 SER 471 PART NUMBER MC UM ARC DATE	T/L QTY EXT	STD QTY	STD COST	STD COST EXTENDED	MATL COST	MATERIAL COST EXT	SU	RT EXT A/C	LABOR EXT D	UNIT COST	UNIT C EXT.-NHA	PAGE 1799
757-0230-001 08 07 114554 099	1.0	200	1.07	1.07	.14	.14	1.8	.07 33	.4		5224653001	
757-0230-001 85 07 114554	1.0										7570230001	
757-8953-001 01 07 117000 039	1.0	200	.60	.60			1.8	.04 32	.2	.60	.60	
757-8954-001 01 07 115004 079	1.0	400	1.02	1.02	.01	.01	2.5	.08 33	.4			
757-8955-001 01 07 116601 039	1.0	300	.32	.32	.32	.32				.32	.32	
757-8956-001 01 07 115900 019	1.0	2000	.62	.62	.62	.62				.62	.62	
757-8959-001 01 07 116600 048	1.0	171	.03	.03	.03	.03				.03	.03	
757-8960-001 05 07	1.0											
UNPRICED PART NUMBER												
763-9265-001 01 07 211214 079	1.0	80	4.90	4.90								
763-9266-001 81 07 124204 078	1.0	50	1.60	1.60	.03	.03	2.2	.07 33	.6		7639269001	
763-9267-001 01 07 114204 038	1.0	175	.76	.76	.02	.02	1.7	.05 33	.3	.76	.76	
763-9268-001 01 07 115003 069	1.0	55	1.64	1.64	.02	.02	2.1	.09 33	.6			
763-9269-001 85 07 114252	1.0										7639269001	
763-9269-001 08 07 114252 069	1.0	160	2.64	2.64	1.89	1.89	1.5	.05 33	.3		5224653001	
763-9270-001 05 07	1.0											
UNPRICED PART NUMBER												
763-9271-001 05 07	1.0											
UNPRICED PART NUMBER												
763-9275-001 01 07 114201 019	1.0	200	.13	.13			.7	.01 33	.1	.13	.13	
767-0920-000 04 07 210090 029	1.0		.27	.27						.27	.27	
TOTALS--ARC ONE			12.01		4.66	14.3		.46	2.9	3.95	PUR/AG	
TOTALS--ARC TWO+FOUR			23.19							1.93	.81 PUR/AG	
TOTALS--ARC FIVE											PUR/AG	

Figure 4. Estimating Tabulation.

and manufacturing lead time (and this can be accomplished by clerical personnel) the bid is produced with a minimum of effort by our skilled estimators. I might add that there is presently an irritating amount of clerical manipulation required to delete excess information. We hope to get the program modified to reduce this "information explosion" to a more acceptable level.

Leaving that simplest situation, let's suppose that we want to bid a new equipment that is composed of a mixture of existing parts and new parts. The designer again marries the base files via the estimating applications program. The existing part numbers play in as before; but new part numbers have no history to fall back on. We're not dead though--with as many different products and parts as we have, a new device has just got to be similar to something that already exists, and the designer may reference the "similar to" item, again helping our cause. If worse comes to worse, and the design has not progressed to this degree of definition, the designer can at least input a dummy part number and description, thereby minimizing the possibility of sins of omission in the bill of material or in our rendering of the estimate. In this situation, the power of the cost retrieval system allows the skilled estimator to devote his time to the new parts and communication with the designer, rather than rehashing old business. Not everyone has a need for a system with this kind of sophistication, but everyone can benefit from a fast, accurate retrieval system geared to his particular size and complexity.

Principle 4. <u>Treat detailed cost estimating as separate from pricing activities.</u> Notice that we have been dealing with the subject of estimating at the standard cost or manufacturing cost level. The reason is this. It is our feeling that the basic cost estimate must be arrived at as <u>objectively</u> and <u>consistently</u> as possible, and that the estimator should not get involved in the business of bid adjustments which consider profit markup, overhead rate predictions, competition, and the like. Now, I do not mean that he shouldn't be involved in cost reduction activities; I mean that once the cost reduction changes are arrived at, he should render the new cost prediction objectively, and leave the rest to the financial and marketing folks.

In this fashion, you create a real solid bench mark from which to manage your bid strategy. Failure to do so only complicates the work of the marketing strategists, and they tell me they don't need any more excitement in their lives.

Principle 5. <u>Measure estimate schedule, quality, and cost.</u> I think most people have some indicators of estimate timeliness, and quality. Keeping score on timeliness is a relatively easy matter--just have someone jot down how many estimates are rendered on time and how many are not. In our case, we do keep score ourselves, but with the cost estimate coordinator serving as our conscience, it is almost impossible for us to be behind schedule and not know it. Measuring quality is not quite so straightforward due to elapsed time, and once again. the many changes that can occur between the time of the estimate, and the time of performance. Design changes, quantity changes, method changes can all confound the evaluation of estimate accuracy. This is particularly true in a business where technological change is a way of life. Still, it is essential to regularly compare the estimate to the actual, or in our situation, to the standard cost at the time of performance. This may be done on a sample basis, and despite the problems I have mentioned, can give you a good feel for trends, or, a managment by exception opportunity if the indicator goes outside your concept of control limits.

Figure 5 shows six consecutive months and a twelve-month average extracted from a controller's report, which serves as our primary quality control indicator. The total region's ratio of cost estimate to standard is on the left; and our area of responsibility, fabrication division, is on the right. We feel that we are in good control, and I use these statistics like a drunk uses a light post--more for support than illumination.

	TOTAL STANDARDS	FABRICATION	ELECTRICAL PARTS	LABOR
4 WEEKS ENDED 9-27-68	97.70	98.10	99.44	99.65
5 WEEKS ENDED 11-1-68	96.67	99.49	96.22	100.52
4 WEEKS ENDED 11-29-68	97.80	97.17	101.53	94.69
4 WEEKS ENDED 12-27-68	98.79	96.24	103.20	94.73
5 WEEKS ENDED 1-31-69	99.20	99.22	103.50	93.52
4 WEEKS ENDED 2-28-69	100.80	101.30	105.13	94.93
12 MONTHS AVERAGE 3-2-68 THROUGH 2-28-69	98.63	99.56	100.33	96.15

Figure 5. Performance Index on Cost Estimating.

Phil Ostwald has handled the subject of cost so well that further discussion here is unnecessary.

Principle 6. <u>Provide</u> <u>for</u> <u>reviews</u> <u>of</u> <u>estimates</u> <u>with</u> <u>various</u> <u>management</u> <u>levels</u>. There are a number of objectives to be served by observing this principle. We review the estimate on "significant" parts with the designer on most equipments. The "significant" part concept allows us to improve our objective of quality, by catching design input errors and subsequent processing errors without wasting a lot of estimator and designer time on all the parts in an equipment. The phenomenon that makes this possible is the fact that on most of our equipments, 10 to 15 percent of the parts content accounts for 80 to 90 percent of the cost. This 10 to 15 percent of the parts then, are the "significant" parts.

Another objective of the "significant" parts review is to bring to the designers attention those parts which have the most effect on the cost of the product, and therefore, the most potential for simplification and cost reduction. Reviews are occasionally held with our production, production control, and quality control managers. This is usually done when the equipment has some new or unusual contractual requirements like extra QC documentation, or when the drawing conventions used are foreign to the shop. They are then in a position to help us evaluate the cost impact of unusual requirements. They also do a pretty good job of grilling us about how the estimate was generated, and "did we use appropriate allowances for scrap, status of design, and lead time factors," as well as our concept of release quantities on jobs which will run over a long calendar period.

Reviews above this level are generally organized and scheduled by the cost estimate coordinator, where more "reasonableness" tests are made, comparing the current estimate to past estimates and to similar equipment estimates; testing the ratio of mechanical to electrical to assembly to total dollars to see whether there is some unexplainable phenomena in the estimate. At this level, we are free to consider the pricing strategy once we are sure the bench mark estimate is sound.

For purposes of summary, I would like to put these six principles in the context of the primary management tasks of planning, organizing, and controlling.

Planning, organizing, and controlling are common to all managed activities from machining to moon shots, and from homemaking to homebuilding. Yes, even the homemaker, if she's a good manager, will plan your weekend activities, give you an organized list of things you should do, and then exercise control to the point where you need to go to work on Monday to rest up.

Which principle could logically come under the head of planning? Well, certainly the principle of documenting procedures, formulas, estimating data, and responsibilities must represent your plan--how you expect things to be done. As was pointed out, your plan may be unique to your business, but the failure to document that plan and have it well understood could be the end of your business.

How about organizing? We talked about three principles which I feel are best categorized here. Appointing only experienced, qualified people as estimators is a part of organizing. Your system for rapid, accurate retrieval of cost information is, hopefully, a model of organized data and activity. And the principle of treating cost estimating as separate from pricing is an organizational way of promoting objectivity and consistency in your estimates.

Observing the principles, which I have grouped under organizing, then, I believe creates a situation where we have a reason to expect the realization of that solid bench mark which we want an estimate to be. But, even having done these good things, success can't be taken for granted.

And that brings us to controlling. In order to be sure that the plan is sound and that the organization is performing according to the plan, we have suggested controlling by the measurement of estimate schedule, quality and cost, and the principle of management review to insure that any trends away from your plan are detected and corrected.

Well there it is; we didn't stop marketings' age-old romance with the customer--we didn't get the designer to invent any sooner or stop revising, and we didn't tell the customer his bid opening date is ridiculous. What we did, I hope, is to point to some principles that will help in doing a successful job despite the challenging environment.

INDEX

A

Absorption costing, 131, 136
Accuracy, 115
Activity Profiling, 83-85
Advertising, 121
Alpha Pressure Bottle, 98-99, 102-103
AMPS, See: Automated Manufacturing Planning System
Angular baseplates, 193
Assembly, 245-252
Attendance Hours, 11
Audit path, 115, 118
Authors
    Berman, Robert, 237-244
    Burch, Jim, 223-226
    Cyrol, E. A., 60-67
    Dewelt, Robert, 217-222
    Doney, Lloyd, 71-75
    Field, Michael, 15-28, 142-149
    Fowler, Paul, 15-28, 142-149
    Friedman, Moshe, 15-28
    Fruehwirth, Michael Z., 35-37
    Gantz, Gordon, 9-14
    Johnson, John E., 104-108
    Jordan, Raymond B., 167-182
    Kriegler, Arnold M., 245-252
    Lindgren, Leroy H., 45-59
    Malolepszy, Adam, 38-44
    Margets, J. G., 123-125
    Murphy, Romeyn D., 45-52
    Nelson, Leonard, 185-196
    Ostwald, Phillip, 29-32, 38-44, 111-122, 129-141, 150-154, 197-201, 205-208, 214-216, 229-232
    Paul, A. N. 3-8
    Radke, Donald G., 76-79
    Strohecker, Daniel, 90-103
    Tipnis, Vijay, 15-28
    Toole, Patrick, 29-32, 197-201
    Von Kaas, H. K., 209-213
    Wenig, Raymond, 80-89
    White, Gary, 129-141, 214-216
    Young, Samuel, 126-128
Automated Manufacturing Planning System, 71-75

B

Base Plates
    angular, 193
    bench fixtures, 119, 192
    five locators, 194
    special, 193
    time values, 192
    turning, 192
    two operations, 193
Basic Blanking Dies, 187-188
Basic Extruding Dies, 188
Basic Hole Piercing Dies, 186-187
Batelle Memorial Institute, 90
Bench fixtures, 191, 192
Bending, 191, 192
Bending dies, 188-189
Boeing Concept, 181
Boeing Northrup Problem, 180-181
Break-even Analysis, 34
Burden, 7, 9
Burden cost, 219
Burden rates, 123-125
Burring, 191
Business interruptions, 179
Buy or Make decision, See: Make vs buy

C

CAM, See: Computer Aided Manufacturing
Capacity factor, 205
Capacity planning, 46
Capture percentage, 115, 118
Clamps, 196
Clocked hours, 14
Coincidence factor, 205
Combination dies, 189-190
Companion estimates, 115
Company losses, 7
Comparison Method, 118, 153
Composite rates, 15
Computer Aided Manufacturing, 15-28
Computer applications, 15-28, 45-59, 71-75, 76-79, 80-89, 90-103, 104-108
Computerized Standard Data, 45-59
Computerized Standard Time, 45-59
Conference Method, 153
Confidence, 243
Constant Slope Myth, 176
Consumer products, 115
Continuous Production Logging, 81-82
Contracted costs, 31
Contract estimating, 29
Contributed value, 9
Converting type joint costs, 41
Cost
    absorption, 131, 136
    accounting, 3
    assembly, 245-252
    burden, 123-125, 219
    contractural, 31
    control, 129-141
    current, 31
    delivery, 31
    direct, 6, 129, 131
    direct labor, 6, 7, 9, 10, 11, 123, 129, 135, 197-201, 211, 123, 229, 231
    durable goods, 209-213
    elements of, 231
    equipment, 34, 121
    fabrication, 245-252
    factory, 10, 11
    fixed indirect, 131
    follow-on, 159
    four-slides, 33-34
    full variable, 116, 118
    future, 241
    historical, 86-87, 140
    indirect labor, 7, 123, 137, 211, 213
    inspection, 35-37, 236
    job, 10
    job shops, 116, 118
    joint, 38-44, 116, 118
    labor, 33, 34, 113, 130, 134, 219, 246
    last, 31
    lead time replacement, 31
    management, 88-89
    marginal, 129
    materials, 7, 34, 197-201, 218-219, 236
    molding, 104-108
    operating control, 10
    original, 31
    out-of-pocket, 129, 135
    past average, 5
    plastic parts, 104-108
    point-in-time, 31
    prime, 10, 136
    product, 12, 129-141, 222
    quotation, 31, 32
    realized, 31
    reduction, 115
    repair, 35, 37
    reported, 11
    set up, 219-220
    stamping, 33-34
    standard, 11, 217-218, 220
    stock ordered, 10
    tooling, 33-34, 104-108, 121, 135, 137, 197-201, 231, 246
    total, 10, 33-34
    total factory, 9
    true, 32
    true inventory, 31
    unit, 30, 32, 113
    woodgrain plastic parts, 104-108

Cost control, 129–141
Cost elements, 231
Cost Estimate Input Sheet, 78
Cost Estimating Committee, 231
Cost Estimating Department, 231
Cost histories, 86–87
Cost of Goods Manufactured, 134
Cost reduction, 115
CSD, See: Computerized Standard Data
Cumulative curve, 163–164
Current cost, 31
Cutting fluids, 15
Cutting rates, 27–28
Cutting tools, 25

D

Daily Status Report, 247
Data and Design Parameters, 150
Data logging, 81–83
Day Work Hours, 11, 14
Decentralized cost estimating, 236
Defective work, 83
Degree day, 206
Delivery cost, 31
Demand factors, 205
Department Cost Method, 114, 117
Design changes, 177–179
Design-to-cost, 129, 134–135
Desperation estimating, 115
Details Developed Quote, 147
Developmental work, 142–149
Dies
    basic hole piercing, 186–187
    bending, 188–189
    blanking, 187–188
    combination, 189–190
    extruding, 188
    forming, 188–189
    manufacturing expenses, 137
    progressive, 190
Direct cost, 6, 131
Direct labor, 6, 7, 9, 10, 11, 123, 129, 135, 197–201, 211, 213, 229, 231
Direct material, 7, 10, 123
Discrete Production Logging, 82–83
Distributed Cost Management, 88–89
Distributing, 39–40
Diversity factor, 205
Divisional expense, 124
Drilling, 191
Durable goods, 209–213
Duration curve, 205
DWL, See: Day Work Hours

E

Earned Productive Hours, 11
Earned Standard Hours, 11
Ease of preparation, 116
Electronic circuit boards, 88
Energy, 82, 84–85, 205–208
Energy losses, 205
Engineering-Change Orders, 166
Engineering costs, 3
EPH, See: Earned Productive Hours
Equipment breakdowns, 83
Equipment costs, 34, 121
Estimated Learning Slope, 122
Estimating form, 132, 138
Estimator
    age, 229
    company size, 230
    education, 229, 233
    number, 230
    responsibility, 230, 234, 246
Expense accounts, 151
"Experience Based Quote," 147
External products, 143
External Reporting, 136
Extruding dies, 188

F

Fabrication, 145–252
Factor Method, 152, 154
Factory cost, 10
Factory cost dollars, 11
Factory Excess Ratio, 210, 211
Factory Operating Statement, 211
Factory overhead, 7, 9
Factory Support Ratio, 210
Federal Procurement Regulation, 116, 118
Fixtures
    bench, 191
    charges, 7
    drilling, 191
    estimating errors, 8
    milling, 191
    physical limits, 191
    turning, 191
Follow-on-costs, 159
Form 633, 116
Forming dies, 188–189
Four-slides, 33–34
Full variable cost, 116, 118
Future costs, 241

G

Guide bushings, 196

H

Heat Factor, 206
Historical cost estimating, 86–87, 130, 140

I

Imprecise environment, 219
Incentive, 130
Incentive Coverage, 210
Incentive Gains, 210
Indirect Labor, 7, 123, 137, 211, 213
Industrial products, 116
Industrial Revolution, 130
Inner Comparison Factor, 112–113
Inspection, 35–37, 236
Intensity, 116
Internal Cost Analysis, 130
Internal products, 143
Internal Revenue Service, 130, 136, 201
Inventories, 135, 136
IRS, See: Internal Revenue Service

J

Jigs,
    boring, 196
    charges, 7
    elements of estimate, 6
    estimating errors, 8
    physical limits, 191
Job costs, 10
Job Labor Performance, 13
Job operations, 10
Job parts, 10
Job Productivity Index, 251
Job Scheduling Factor, 13
Job Performance, 14
Job Routing File, 51
Job shops, 116, 118
Job variances, 12
Joint costs, 38–44, 116, 118
Joint products, 38

K

Kilovolt-ampere, 205
Kilowatt, 205
Kilowatt-hour, 205

L

Labor cost, 33, 34, 113, 130, 134, 219, 246
Labor Cost Performance, 13
Labor estimate, 133, 138
Labor rate, 87, 123-125
Last cost, 31
Lathes, 191
LCC, See: Life Cycle Costing
Lead time replacement costs, 31
Learning curve
    business interpretation, 178
    constant slope myth, 176
    cost elements, 176
    cost histories, 86
    cumulative curve, 163-164
    defined, 157-158
    engineering change orders, 166
    follow-on-cost, 159
    least-square fit, 164-165
    log-log graph, 158, 169-170
    proforma, 114, 118
    "S" shaped curve, 177
    specific cost application, 165-166
    theory, 159-162, 168
    use of technique, 167-168
Least-Squares Fit, 164-165
Life Cycle Costing, 214-216
Line Item Estimate, 153-154
List Price, 6, 141
Load factor, 205
Load hours, 11
Locators, 194
Log-log graph, 158, 169-170
Loss factor, 206

M

Machinability data, 15-28
Machinability Data Bank, 15-16
Machinability Data File, 16
Machine selection, 71
Maintainability, 146
Maintenance Transaction Cards, 59
Make vs buy, 116, 118, 236
Management Information Systems, 51, 52
Manufacturing estimating, 29
Marginal Propensity, 112, 116, 118
Mass transportation, 85
Material cost, 7, 34, 197-201, 218-219, 236
Material estimating, 17-32, 34, 138
MDF, See: Machinability Data File
Methods and Standards Generation Program, 47-48
Methods File Maintenance, 48-49
Milling, 15-28, 191
Minicomputers, 80-89
Model Building, 16
Molding, 104-108
Money-out-of-pocket, 31, 32
Motivation, 142-149
Mounting plates, 191-192
Multiple phases of product life, 172-176

N

National Numerical Machining Data Bank, 15
NC, See: Numerical Control
New Products, 150-154
Northrop Concept, 181
Numerical Control, 15-28

O

Old Products, 150-154
150-Unit Cumulative Cost, 121
150-Unit Cumulative Factor, 121
Operating cost control, 10
Operating estimates, 134
Operation Method, 118
Optimization, 16
Original costs, 31
Overall ratios, 5
Overbidding, 8
Overhead, 7, 10, 33, 113, 129, 131-132, 136, 141, 197-201
Overlay Cycle Time, 108
Out-of-pocket cost, 129, 135

P

Packaging, 78, 236
Part Handling, 78
Past average cost, 5
Performance Index, See: Job Performance Index
Period of use, 206
Personnel, 4, 140
PI, See: Job Productivity Index
Planning, 142-149
Plastic parts, 104-108
Point-in-time costs, 31
Predetermined time standards, 46
Preliminary Economic Evaluation, 150
Preliminary estimate, 4-5
Presses, 78
Price, 6, 141
Price-in-effect, 31, 32
Prime cost, 10, 136
Process planning, 133
Process routing, 71
Process Screening Quality Level, 35-37
Procurement, 236
Product costing, 121, 129-141, 222
Production planning, 46
Production rate, 18-20
Productivity, 146
Profit, 6, 131, 134, 135, 141, 150, 212-213
Profit planning, 129-141
Proforma, 111-122, 231
Progressive dies, 190
Project analysis, 130
Project unit costs, 30-37
Projection, 239-240
Propellant rocket motor cases, 90-103
Proprietory product, 117
Prototypes, 150
PSQL, See: Process Screening Quality Level

Q

Quality assurance, 236
Quality control, 121
Quality variances, 83
Quick response estimates, 4
Quotation costs, 31, 32
Quotations, 31, 32, 146-147
Quote Analysis, 125

R

Rate-Tool life, 17-18, 24
Raw material, 82, 121, 132, 246
Rawterial, 44-45
Reactive kilovolt-ampere, 206
Realized costs, 31
Re-estimate, 117, 118
Relative design complexity, 146
Reliability, 146
Repair costs, 35-37
Reported costs, 11
Routing, 11
Royalties, 121

S

Sales expense, 10
Securities and Exchange Commission, 136
Selling expense, 6
Selling price, 6
Service life, 146
Setup costs, 219-220
Setup hours, 113
Setup identification, 117
SH, See: Standard Hour
Shop Methods File, 51
SIC, See: Standard Industrial Classification
Software Packages, 45, 59
Spare parts, 117, 118
Stamping, 33-34
Standard Costs, 11, 12, 217-218, 220
Standard Cost Systems, 217-218, 220
Standard Data, 8, 46
Standard Earnings, 121
Standard Fixed Burden Rate, 131
Standard Hour, 11
Standard Industrial Classification, 229, 233
Standard Time Data, 60-68, 130, 140, 231, 236
Steam, 206
Stock ordered cost, 10

Strikes, 17
Surveys, 229-232

T

Task analysis, 87
Time study, 46, 62
Time study technicians, 229
Tool life, 16, 17
Tooling costs, 33-34, 104-108, 121, 135, 137, 197-201, 231, 246
Tooling design, 133
Tooling Estimate Log, 133, 139
Tools, 6, 7, 8, 134, 137, 140
Top Management, 237-244
Total cost, 10, 33-34
Total factory costs, 9
True Cost, 32
True Inventory Costs, 31
Tube bending, 192
Turning fixtures, 191
Turret lathes, 65, 191

U

Unemployment, 135
Unit costs estimates, 32, 113
Unit Pricing Computation, 134, 139
U.S. Army Missile Command, 90
Utilization Factor, 206

V

Value engineering, 130
Variable Budget, 221
Variable Cost Estimating, 129-141
Varible Cost Method, 114, 117
Variable Manufacturing Costs, 121
Variable Marketing Costs, 121
Variable overhead, 129, 133, 134, 135
Versatility, 118
Volt, 206
Voltage drop, 206
Volume sensitivity, 117, 118
Volume variance, 131

W

Water power, 206
Woodgrain plastic parts, 104-108
Worker absence, 83
Worker control, 86